ASTRAL
AF653144

Advances in Aquatic Ecology

— Volume 5 —

Editor
Dr. Vishwas B. Sakhare
Head,
Post Graduate Department of Zoology
Yogeshwari Mahavidyalaya,
Ambajogai – 431 517
Maharashtra

2011
DAYA PUBLISHING HOUSE®
Delhi - 110 035

© 2011 VISHWAS BALASAHEB SAKHARE (1974–)
ISBN 81-7035-697-0
ISBN 978-81-7035-697-4

All rights reserved. Including the right to translate or to reproduce this book or parts thereof except for brief quotations in critical reviews.

The views expressed in various articles are those of the authors and not of editor or publisher of the book.

Published by : **Daya Publishing House®**
4760-61/23, Ansari Road, Darya Ganj,
New Delhi - 110 002
Phone: 23245578, 23244987
Fax: (011) 23260116
E-mail: dayabooks@vsnl.com
website: www.dayabooks.com

Laser Typesetting : **Classic Computer Services**
Delhi - 110 035

Printed at : **Chawla Offset Printers**
Delhi - 110 052

PRINTED IN INDIA

Advances in Aquatic Ecology

— Volume 5 —

Preface

Not only scientists, but common man and rulers are also worried about the increasing threats to our aquatic ecosystems. Therefore basic knowledge about the aquatic ecosystems, pollution and conservation is essential. In this context *'Advances in Aquatic Ecology* (*Vol. 5*)' focuses the issues to preserve these precious ecosystems.

This volume is a unique compilation of 42 research papers which discusses exhaustive studies on physico-chemical environment, planktonology, fisheries, toxicology etc. With its application oriented and interdisciplinary approach, the book would be immensely useful to students, teachers, researchers, scientists, policy makers, environmental lawyers and others interested in aquatic ecology.

The chapters in the book have been contributed by eminent scientists/academicians active in the areas of aquatic ecology. I am thankful to Dr. Sureshji Khursale, President, Yogeshwari Education Society, Dr. P.S. Prayag, Principal, Yogeshwari Mahavidyalaya, Ambajogai and all staff members of Yogeshwari Mahavidyalaya, Ambajogai for their encouragement. The co-operation of my departmental colleagues and research scholars is also thankfully acknowledged. Special thanks are due to Mr. Anil Mittal of Daya Publishing House, Delhi for taking pains in bringing out this volume. I take this opportunity to thank everybody who had helped me in preparation of this volume.

Last but not least, I am grateful to my teacher and research guide Dr. P.K. Joshi, Reader in Zoology, Dnyanopasak College, Parbhani for his blessings. I am also thankful to Dr. Patricio De los Rios of Catholic University of Temuco (Chile), Dr. Indranil Ghosh of West Bengal University of Animal Sciences and Fisheries, Kolkata, Dr. S.S. Todkari and M.M. Girkar of College of Fisheries, Udgir, Dr. B.R. Chavan of College of Fisheries, Ratnagiri, Dr. Meenakshi Jindal of C.C.S. Haryana Agricultural University, Hisar (Haryana), Dr. Sajid Maqsood of Sher-e-Kashmir University of Agricultural Science and Technology, Kashmir, Dr. V.S. Shembekar of Rajarshi Shahu College, Latur, Dr. Syeda Azeem Unnisa of Jawaharlal Nehru Technological University, Kukatpally-Hyderabad, Dr. C. Stella and J. Siva of Alagappa University, Thondi (Tamil Nadu), Dr. V.R. Borane of Jijamata Arts, Science, and Commerce College, Nandurbar (Maharashtra), Dr. N.V. Harney, Dr. S.R. Sitre and Dr. P.N. Nasare of Nilkanthrao

Shinde Science and Arts College, Bhadrawati (Maharashtra), Dr. R.P. Mali of Yeshwant College, Nanded, Dr. K. Kannan and G. Rajasekharan of Government Arts College, Ariyalur, Dr. Chand Pasha of Osmania University, Hyderabad, Dr. Ashok Mohekar and Dr. D.A. Kulkarni of Shikshan Maharshi Dnyandev Mohekar College, Kallam (Maharashtra), Dr. Muthumurugan of Ayya Nadar Janaki Ammal College, Sivakasi, Dr. S.A. Khabade of D.K.A.S.C. College, Ichalkaranji (Maharashtra), Dr. M.B. Mule and Dr. Ram Chavan of Dr. Babasaheb Ambedkar Marathwada University, Aurangabad, Dr. A.S. Kulkarni and M.V. Tendulkar of R.P. Gogate College of Arts and Science and R.V. Jogalekar College of Commerce, Ratnagiri, Prof. M. Bhaskar of S.V. University, Tirupati (Andhra Pradesh), Dr. D.V. Muley of Shivaji University, Kolhapur, Dr. C.M. Bharambe of Vidnyan Mahavidyalaya, Malkapur (Maharashtra), Dr. Shyam S. Salim of Central Institute of Fisheries Education, Mumbai and Dr. D.P. Singh of S.M.L. (P.G.) College, Jhunjhunu (Rajasthan) and Dr. L.P. Lanka of Devchand College, Arjunanagar (Maharashtra) for giving me encouragement and valuable suggestions.

Dr. Vishwas B. Sakhare

Contents

List of Contributors

Acevedo, Patricio
Universidad de la Frontera, Facultad de Ingeniería y Administración, Departamento de Ciencias Físicas, Casilla 15-D, Temuco, Chile

Ashashree, H.M.
Sahyadri Science College (Autonomous), Shivamogga, Karnataka

Balasubramanian, V.
Post-graduate and Research Department of Zoology, Ayya Nadar Janaki Ammal College (Autonomous), Sivakasi – 626 124

Balkhi, M.H.
Faculty of Fisheries, Sher-e-Kashmir University of Agricultural Sciences and Technology, Kashmir

Barve, S.K.
College of Fisheries, Shirgaon, Ratnagiri – 415 612

Bharambe, C.M.
Department of Zoology, Vidnyan Mahavidyalaya, Malkapur – 443 101

Bhaskar, M.
Department of Zoology, S.V. University, Tirupati – 517 502

Biswas, P.
Department of Fisheries, I.G.K.V., Raipur

Borane, V.R.
Jijamata Arts, Science, and Commerce College, Nandurbar – 425 412

Chalak, Ashwini D.
Post Graduate Department of Zoology, Yogeshwari Mahavidyalaya, Ambajogai – 431 517

Chaouhan, Vinod Kumar
Department of Microbiology, Nizam College, Osmania University, Hyderabad – 7

Chati, R.S.
Department of Zoology, Karmveer Mamasaheb Jagdhale Mahavidyalaya, Washi – 413 503

Chaudhari, K.J.
College of Fisheries, Shirgaon, Ratnagiri – 415 612

Chavan, B.R.
College of Fisheries, Shirgaon, Ratnagiri – 415 629

Dahiya, T.P.
Department of Zoology and Aquaculture, CCS Haryana Agricultural University, Hisar – 125 004

Dar, Shabir Ahmad
Central Institute of Fisheries Education, Versova, Mumbai – 400 061

Deepthi, P.
Centre for Environment IST, Jawaharlal Nehru Technological University, Kukatpally, Hyderabad – 85

Deshmukh, A.V.
College of Fisheries, Shirgaon, Ratnagiri – 415 612

Dhamagaye, H.B.
Marine Biological Research Station, Ratnagiri – 415 629

Dhavle, S.D.
Department of Botany, S.A.S. College, Mukhed – 431 715

Funde, A.B.
College of Fisheries, Shirgaon, Ratnagiri – 415 612

Gawde, M.M.
College of Fisheries, Shirgaon, Ratnagiri – 415 612

Ghosh, A.
Department of Fisheries, I.G.K.V., Raipur

Girkar, M.M.
College of Fisheries, Shirgaon, Ratnagiri – 415 612

Hakeem, M.D.
Department of Microbiology, Nizam College, Osmania University, Hyderabad – 7

Harney, N.V.
Department of Zoology, Nilkanthrao Shinde Science and Arts College, Bhadrawati – 442 902

Hauenstein, Enrique
Universidad Católica de Temuco, Facultad de Recursos Naturales, Escuela de Ciencias Ambentales, Casilla 15-D, Temuco, Chile

Indulkar, S.T.
College of Fisheries, Shirgaon, Ratnagiri – 415 629

Injal, A.S.
Department of Zoology, R.P. Gogate College of Arts and science and R.V. Jogalekar College of Commerce, Ratnagiri – 415 612

Jadhav, B.V.
Department of Zoology, Dr. B.A.M. University, Aurangabad – 431 004

Jadhav, V.S.
Department of Zoology, V.N. College, Vasarni, Nanded – 431 605

Jalal, Firozia Naseema
P.G. and Research Department of Zoology, N.S.S. College, Pandalam – 689 501

Jaque, Ximena
Universidad Católica de Temuco, Facultad de Recursos Naturales, Escuela de Ciencias Ambentales, Casilla 15-D, Temuco, Chile

Jindal, Meenakshi
Department of Zoology and Aquaculture, CCS Haryana Agricultural University, Hisar – 125 004

Joshi, P.K.
Dnyanopasak Mahavidyalaya, Parbhani – 431 401

Joshi, R.D.
Department of Microbiology Yogeshwari Mahavidyalaya, Ambajogai – 431 517, Maharashtra

Kale, M.D.
Department of Zoology, Shivaji University, Kolhapur – 416 004

Kannan, K.
Department of Zoology, Govt. Arts College, Ariyalur – 621 713

Kanwate, V.S.
Department of Zoology, V.N. College, Vasarni, Nanded – 431 605

Khabade, S.A.
Department of Zoology, D.K.A.S.C. College, Ichalkaranji, Maharashtra

Khobragade, Kshama
S.B.E.S. College of Science, Aurangabad – 431 001

Kulkarni, A.S.
Department of Zoology, R.P. Gogate College of Arts and science and R.V. Jogalekar College of Commerce, Ratnagiri – 4156 12

Kulkarni, D.A.
P.G. Department of Zoology, Shikshan Maharshi Dnyandev Mohekar College, Kallam – 413 507

Kulkarni, G.N.
College of Fisheries, Shirgaon, Ratnagiri – 415 612

Kumar, Sudhir
Department of Biotechnology, Goa University, Goa – 403 206

Kumari, Switi
Padmashree Dr. D.Y. Patil A.C.S College, Akurdi, Pune – 44

Lakde, H.M.
Department of Botany, Gramin College, Vasant Nagar, District Nanded

Lanka, L.P.
Department of Zoology, Devchand College, Arjunanagar

Lohare, S.D.
Department of Botany, H.S. College, Udgir – 413 517

Madhuri, E.
Department of Zoology, S.V. University, Tirupati – 517 502

Madlapure, V.R.
Gramin Mahavidyalaya, Vasantnagar Kotgyal, Tq. Mukhed District Nanded

Mali, R.P.
P.G Department of Zoology, Yeshwant Mahavidyalaya, Nanded – 431 602

Maqsood, Sajid
Faculty of Fisheries, Sher-e-Kashmir University of Agricultural Science and Technology, Kashmir

Mohekar, A.D.
Department of Zoology, S.M.D.M. College, Kallam – 413 507

Mohite, J.S.
Yeshwantrao Chavan Mahavidyalaya, Tuljapur – 413 601

Mohite, S.A.
College of Fisheries, Shirgaon, Ratnagiri – 415 612

Mukkanti, K.
Centre for Pharmaceutical Sciences, IST, Jawaharlal Nehru Technological University, Kukatpally, Hyderabad – 85

Mule, M.B.
Department of Environmental Science, Dr. Babasaheb Ambedkar Marathwada University, Aurangabad – 431004

Muley, D.V.
Department of Zoology, Shivaji University, Kolhapur – 416 004

Murugan, A.
Gulf of Mannar Biosphere Reserve Trust, Ramanathapuram

Muthumurugan, N.
Post-Graduate and Research Department of Zoology, Ayya Nadar Janaki Ammal College (Autonomous), Sivakasi – 626 124

Naik, S.D.
College of Fisheries, Shirgaon, Ratnagiri – 415 629

Najar, A.M.
Sher-e-Kashmir University of Agricultural Sciences and Technology of Kashmir, Faculty of Fisheries, Ganderbal, Kashmir

Nakhawa, A.D.
College of Fisheries, Ratnagiri – 415 629

Nasare, P.N.
Department of Botany, Nilkanthrao Shinde Science and Arts College, Bhadrawati – 442 902

Nikam, S.M.
Department of Zoology, R.P. Gogate College of Arts and science and R.V. Jogalekar College of Commerce, Ratnagiri – 415 612

Pai, R.
College of Fisheries, Shirgaon, Ratnagiri – 415 629

Pasha, Chand
Department of Microbiology, University College of Science, Osmania University, Hyderabad

Patil, R.D.
Arts, Science and Commerce College, Navapur – 425 418

Patil, S.R.
Department of Zoology, Y.C. College, Warnanagar

Patil, S.S.
Department of Environmental Science, Dr. Babasaheb Ambedkar Marathwada University, Aurangabad.

Pavaraj, M.
Post-graduate and Research Department of Zoology, Ayya Nadar Janaki Ammal College (Autonomous), Sivakasi – 626 124

Pawar, A.S.
College of Fisheries, Shirgaon, Ratnagiri – 415 629

Pawar, V. B.
D.S.M. College of Arts, Commerce and Science, Parbhani – 431 401

Pentewar, M.S.
Gramin Mahavidyalaya, Vasantnagar Kotgyal, Tq. Mukhed District Nanded

Rajasekaran, G.
Department of Zoology, Government Arts College, Ariyalur – 621 713

Raje, P.C.
College of Fisheries, Shirgaon, Ratnagiri – 415 612

Rama Krishna, K.S.
Department of Zoology, S.V. University, Tirupati – 517 502

Ramesh Babu
Department of Zoology, S.V. University, Tirupati – 517 502

Rao, J. Venkateswar
Department of Zoology, Nizam College, Osmania University, Hyderabad – 7

Rather, Mohd. Ashraf
Central Institute of Fisheries Education, Versova, Mumbai – 400 061

Reddy, R.M.
Department of Zoology and Fishery Science, Rajarshi Shahu College, Latur – 413 512

Rios, Patricio De los
Universidad Católica de Temuco, Facultad de Recursos Naturales, Escuela de Ciencias Ambentales, Casilla 15-D, Temuco, Chile

Sadafule, N.A.
MFSc (FBM) College of Fisheries, Ratnagiri – 415 629

Sadawarte, R.K.
College of Fisheries, Shirgaon, Ratnagiri – 415 629

Sailaja, V.
Department of Zoology, S.V. University, Tirupati – 517 502

Salim, Shyam S.
Scientist (Senior Scale), Central Institute of Fisheries Education, Mumbai – 61

Samoon, M.H.
Faculty of Fisheries, Sher-e-Kashmir University of Agricultural Science and Technology, Kashmir

Sanal Kumar, M.G.
P.G. and Research Department of Zoology, N.S.S. College, Pandalam – 689 501

Sathe, A.R.
College of Fisheries, Shirgaon, Ratnagiri – 415 629

Shaikh, Afsar
P.G Department of Zoology, Yeshwant Mahavidyalaya, Nanded – 431 602

Sharangdher, S.T.
College of Fisheries, Shirgaon, Ratnagiri – 415 629

Sharma, Rupam
Central Institute of Fisheries Education, Versova, Mumbai – 400 061

Shembekar, V.S.
Department of Zoology and Fishery Science, Rajarshi Shahu College, Latur – 413 512

Shinde, B.R.
Jijamata Arts, Science, and Commerce College, Nandurbar – 425 412

Shirdhankar, M.M.
College of Fisheries, Shirgaon, Ratnagiri – 415 612

Sihag, R.C.
Department of Zoology and Aquaculture, C.C.S. Haryana Agricultural University, Hisar – 125 004

Singh, Anshu
PGT-Biology, DPS, Dundlod, Jhunjhunu – 333 001

Singh, D.P.
P.G. Department of Botany, SML (PG) College, Jhunjhunu – 333 001

Singh, Prabjeet
Faculty of Fisheries, Sher-e-Kashmir University of Agricultural Science and Technology, Kashmir

Sitre, S.R.
Department of Zoology Nilkanthrao Shinde Science and Arts College, Bhadrawati – 442 902

Siva, J.
Department of Oceanography and Coastal Area Studies, Alagappa University, Thondi Campus, Thondi – 623 409

Sri Divya, B.
Department of Microbiology, University College of Science, Osmania University, Hyderabad

Stella, C.
Department of Oceanography and Coastal Area Studies, Alagappa University, Thondi Campus, Thondi – 623 409

Surwase, S.S.
P.G. Department of Zoology, Shikshan Maharshi Dnyandev Mohekar College, Kallam – 413 507

Suvare, V.B.
Department of Zoology, R.P. Gogate College of Arts and science and R.V. Jogalekar College of Commerce, Ratnagiri – 415 612

Tendulkar, M.V.
Department of Zoology, R.P. Gogate College of Arts and Science and R.V. Jogalekar College of Commerce, Ratnagiri – 415 612

Tiwary, Mukesh
Padmashree Dr. D.Y. Patil A.C.S. College, Akurdi, Pune – 44

Todkari, S.S.
College of Fisheries, Ratnagiri – 415 612

Unnisa, Syeda Azeem
Centre for Environment IST, Jawaharlal Nehru Technological University, Kukatpally – Hyderabad – 85

Urmila, Barros
Department of Biotechnology, Goa University, Goa – 403 206

Vardia, H.K.
Department of Fisheries, I.G.K.V, Raipur

Vhanalakar, S.A.
Department of Zoology, Shivaji University, Kolhapur – 416 004

Vijayalakshmi, S.
CAS in Marine Biology, Annamalai University, Parangipettai – 608 502

Wadhave, N.S.
Department of Botany, Nilkanthrao Shinde Science and Arts College, Bhadrawati – 442 902

Waghmode, S.S.
Department of Zoology, Shivaji University, Kolhapur – 416 004

Chapter 1

Agro-based Material for Purifying Turbid Water

☆ *Syeda Azeem Unnisa, P. Deepthi and K. Mukkanti*

Introduction

Groundwater is the preferred source for drinking water in rural areas of developing countries and it generally requires no or minimal treatment. In the event that no suitable aquifers are available, relatively clean waters from lakes or streams are preferred. However, only simple, practical technologies such as gravity chemical feed with solutions, hydraulic rapid mixing and flocculation, horizontal-flow sedimentation, and manually operated filters should be used for treatment of such waters (Schulz and Okun 1984).

Unlike cities where fairly large population is using water filters, Aqua guards, UV–ultra filters but the rural population is thriving on the contaminated water supply due to prohibitive cost and low availability of chemical coagulants and disinfectants and the heavy investment in settling–up the conventional water treatment plants at village level is not only a theoretical exercise but practically impossible for several reasons. Such projections have prompted interest in using traditional methods for treating the water. *Moringa oleifera* seeds treat water on two levels, acting both as a coagulative and as antibacterial agent (Babu and Chaudhuri, 2005). It is generally accepted that Moringa works as a coagulant due to positively charged water soluble proteins, which bind with negatively charged particles (silt, clay, bacteria, toxins etc) allowing the resultant "flocs" to settle to the bottom and can be removed by filtration (Lye, 2002).

Materials and Methods

Sample Collection

Turbid water samples were collected in plastic bottles of 1.5 litres from different lakes *i.e.* Kukatpally lake, Pragathi Nagar lake and Medchal lake, Hyderabad, A.P., India. The physical appearance of samples was mainly turbid in nature and was contaminated by different sources like industrial sewage, municipal sewage and agricultural runoff.

Plant Material

The Moringaceae is a single genus family with 14 known species. Of these *Moringa oleifera* is the most widely known and utilized species (Sutherland *et al.*, 1994). Dried pods of *Moringa oleifera* were brought from Horticulture Nursery Training Centre, Khammam district, A.P., India. Seeds were separated from the pods and then pulverized using a clean pestle and mortar shown in (Figure 1.1).

Figure 1.1: Fine Powder of *Moringa oleifera* Seeds

Coagulant Solution Preparation

The pulverized seed material of about 0.12 gms (optimum) was made into paste using little amount of water and mixed into 5 ml of clean water which was shaked for 1 minute in order to activate the coagulant properties of the seed to form a solution (Jahn, 1986). The solution was filtered through a muslin cloth (to remove insoluble materials) into the 500 ml of turbid water to be treated.

Physico-chemical and Microbial Analysis

Samples were analyzed before and after treatment for its potability using (APHA, 1988 and Cheesbrough, 1984) method. The turbidity of water samples was measured using the turbidity meter (ELICO CL 52 D). The *Escherichia coli* and Coliforms bacteria counts were enumerated on Eosin Methylene Blue and Mac Conkey respectively after 0.1 ml of the turbid water samples was aseptically serial diluted up to three fold.

Methodology

Laboratory trials, carried out at Centre for Environment, IST, JNT University, Hyderabad, India have compared the effectiveness of *Moringa oleifera* for purifying turbid waters. The dried seeds of

Moringa oleifera were pulverized in to fine powder. Small amount of clean water was added to the powder to form a paste. Paste was mixed in to 5 ml of clean water and was shaked for 1 minute to activate the coagulant properties and to form a solution. Solution was filter through a muslin cloth and then poured in to the turbid water. Entire setup was kept for jar test method, where the sample was stirred rapidly for 1 minute and then followed by slow stirring for about 5–10 minutes. Then the samples were kept for 30, 60 and 90 minutes of settling time. After the particles and contaminants have settled to the bottom, the clean water was carefully separated. Then the water was filtered through Watmann filter paper. The water samples were analyzed for its potability before and after treatment with *Moringa oleifera* seeds. All of the experiments were performed in duplicate and the average values were presented.

Results and Discussion

From the visual observation results the turbidity of water samples was high with terrible odor which was then decreased after treatment. The physical nature and microbial load of turbid water samples for the study area are given in Table 1.1. The turbidity reduction was observed in all treated samples when compared with control nevertheless it was high turbid (116 NTU) or low turbid (7.5 NTU) waters which is shown in (Figure 1.2). The turbidity reduction by *Moringa oleifera* seeds were observed in all samples.

Figure 1.2: Comparison of Turbidity Removal *by Moringa oleifera* Seeds

Table 1.1: Turbidity and Microbial Reduction Before and After Treatment with *Moringa oleifera* Seed [optimum dose–0.12 gms, settling time–60 minutes]

Sample Location	*Physical Appearance*		*Turbidity (NTU)*		*Coliform Counts Cfu/ml*		*E. coli Counts Cfu/ml*	
	Initial	*Final*	*Initial*	*Final*	*Initial*	*Final*	*Initial*	*Final*
Medchal	Very dirty, highly turbid with lot of suspended matter	Clear supernatant	116	10.5	TMTC	6000	TMTC	2000
Pragathi Nagar	Foul smell appearing greenish	Clear supernatant	22.5	1.0	TMTC	1000	TMTC	1000
Kukatpally	Rotten foul Smell, appearing greenish	Clear supernatant	7.5	0.75	90,000	500	TMTC	1000

The present study had shown drastic reduction in turbidity and microbial content in the turbid water samples when treated with *Moringa oleifera* seeds. The microbial load (coliforms and *E.coli*) reduction was seen in all samples in 60 minutes of settling time when compared with the other two 30 minutes and 90 minutes. The variance of reduction in different settling times might occurred due to insufficient time for the reaction to complete in 30 minutes settling time and due to disassociation of proteins in 90 minutes settling time. The 60 minutes settling time can be opted as the optimum settling time for treatment as the drastic reduction of turbidity shown in (Figure 1.3) as well as lowest reduction of microbial load was observed which might occurred due to the exact occurrence of the polyelectrolyte

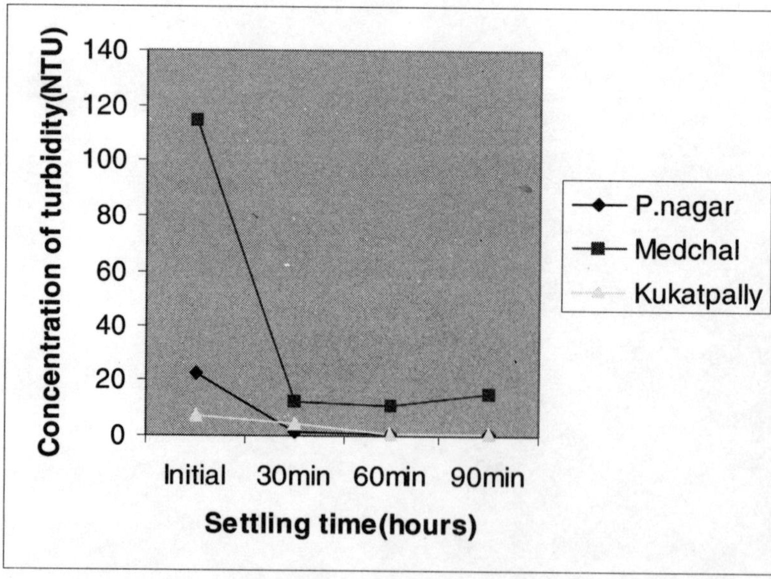

Figure 1.3: Turbidity Reductions with *Moringa oleifera* at Different Settling Intervals

mechanism *i.e.* the seeds of *Moringa oleifera* contain significant quantity of low molecular weight (water soluble proteins), which carry positive charge. When the seeds were crushed and added to raw water, the polysaccharides produce positive charge acting like magnets and attract the predominantly negatively charged particles such as clay, silt, bacteria and toxic particles in water (Mc Connachie *et al.*, 1999). The coagulative property of moringa seeds could be attributed to a polymeric coagulant earlier reported by (Eilert, 1978). The works of Eilert also supports the antibacterial activity of *Moringa oleifera*, while (Umar, 1998) reported the antibacterial action of small protein/peptide against *Escherichia coli, Klebsiella pneumoniae, Staph aureus* and *Bacillus substilis*. The observations in this study corroborate these earlier observations in that *Moringa oleifera* seeds had drastically reduced the *E.coli* and Coliform counts in the turbid water samples. Generally, the turbidity of the turbid water samples reduced drastically to 99 per cent after treating with the *Moringa oleifera* seeds. The pH of the treated and untreated water samples remained at 7.0 which shows that it will not alter the pH of the samples.

Conclusion

The study reveals that the coagulative effects of *Moringa oleifera* had shown results at optimum dose of 0.12 gms and 60 minutes of settling time. The efficiency of treatment is independent of raw water pH and the processing does not modify the pH of the water. Seeds of *Moringa oleifera* contain materials that are effective as coagulant and direct filtration of water with *Moringa oleifera* seed as a coagulant brings about a substantial improvement in its aesthetic and microbiological quality. The positive effect of the *Moringa oleifera* on indicator organisms (*E.coli* and coliforms) means reduction in the level of faecal pollution in the environment. The need to exploit the potential of plants and solar disinfection may offer cheap and environment friendly methods for tackling water contamination and may possibly overcome the hazards of using chlorine. It must be remarked in this study that the *Moringa oleifera* seeds as a water coagulant is gaining wide attention. The study shows an alternative to conventional coagulants, *Moringa oleifera* seeds can be used as a natural coagulant in household water treatment as well as in the community water treatment systems and a very simple and cheap process for rural communities. This process is acceptable for drinking where people are currently drinking untreated and contaminate water.

Acknowledgements

We acknowledge our sincere thanks to the Department of Science and Technology (DST) New Delhi, for sanctioning and granting the financial support for project under the water initiative technology (WTI) programme.

References

American Public Health Association, 1998. *Standard Methods for the Examination of Water and Wastewater*, American Water Works Association and Water Pollution Control Federation, Washington DC., USA.

Babu, R. and Chaudhuri, M., 2005. Home water treatment by direct filtration with natural coagulant. *Journal of Water Health*, 3(1): 27–30.

Cheesbrough, M., 1984. *Medical Laboratory Manual for Tropical Health Technology*. Butterworth, CRC press, Washignton, pp. 1–15.

Eilert, U., 1978. Antibiotic principles of seeds of *Moringa oleifera*. *Indian Medical Journal*, 38(235): 1013–1016.

Jahn, S.A.A., 1986. The tree that purifies water. *Journal of Water Research*, 38: 23–28.

Lye, D.J., 2002. Health risks associated with consumption of untreated water from household roof catchment systems. *Journal of AWRA*, 38(5): 1301–1306.

Umar Dahot, M., 1998. Antimicrobial activity of small protein of *Moringa oleifera* leaves. *Journal of Islamic Academy of Science*, 11(1).

McConnachie, H.L., Folkard, G.K., Mtwali, M.A. and Sutherland, J.P., 1999. Field trial of appropriate flocculation processes. *Journal of Water Research*, 33(6): 1425–1434.

Schulz, C.R. and Okun, D.A., 1984. *Surface Water Treatment for Communities in Developing Countries.* John Wiley and sons, New York.

Sutherland, J.P., Folkard, G.K., Mtawali, M.A. and Grant, W.D., 1994. *Moringa oleifera* as a natural coagulant. In: *Paper Presented at the 20th WEDC Conference*, October 10–15, South Africa.

Chapter 2

Gambusia Fish for Controlling of Pathogenic Water Microorganisms

☆ *Chand Pasha, Vinod Kumar Chaouhan, M.D. Hakeem, B. Sri Divya and J. Venkateswar Rao*

Introduction

Mosquito fish have carried a reputation as the number one biological control for use with mosquitoes (Ghosh and Dash 2007). *Gambusia holbrooki* have a wider distribution throughout the world than any other freshwater fish due to its ability to feed larvae, planktons, eggs of fishes and amphibians and microorganisms (Angler *et al.*, 2002; Boswell *et al.*, 2005). Mosquito fish will prey on new born mosquitoes as well as other small fish. Gambusia feeds about 50 mosquitoes in 30 min in normal pond but in drainage ditches or stagnant water with abundant microbial flora its mosquito feeding is decreased without affecting the normal life (Toft, *et al.*, 2004). The conditions indirectly confirms its preferential feeding on microorganisms rather macrolarvae of mosquito. *G.holbrooki* is widely distributed in both freshwater systems and estuaries of temperate regions. Furthermore, this species is not only abundant but is also easy to capture. In addition, keeping this species under laboratory-controlled conditions is relatively easy. A large number of fish can be reared in small aquaria or tanks, due to the small body size (size between 2 to 2.5 cm long) (Nunes *et al.*, 2008). There is ample of evidence that *Gambusia* poses a threat to endemic species in parts of Australia, New Zealand, and North America. Hence *Gambusia* control strategies are being planned and this should be considered while exploiting the mosquito fish.

The most advanced civilizations hold for water quality, because it represents the first foodstuff of the universe, essential for survival. Water borne issues are the major causative agents for the maximum mortality in threshold. The technologies are available for water decontamination and purification. But they are not being reached to low population villages due to high cost more ever due to ignorance. Human and animal activities in water bodies in villages are responsible for contaminations. In many villages only undergroundwater is being used for drinking. Many reports suggests the contamination of undergroundwater (Vander Grift and Griffioen 2008; Peng *et al.*, 2008). As the undergroundwater levels are reducing and still few villages are depending on water bodies (pond, well and river) for drinking water, there is a need to evaluate and develop methods to control the microbial contamination using the mosquito fish. Hence current study was planned to evaluate the usage of *G.holbrooki* for the control of pathogenic microorganisms in water.

Material and Methods

Mosquito Fish

Mosquito fish *G. holbrooki* was obtained from Dept of Zoology, Nizam College, Osmania University, Hyderabad. Natural growth and breeding is allowed in a small pond, 5 gms of *B. subtilis* powder (10×10^{10} CFU/gm) was added for every 15 days.

Pathogenic Bacteria

Pathogenic bacteria isolated from clinical samples were collected from Vijaya diagnostic center Hyderabad. Bacteria are maintained on nutrient agar slants by sub culturing every 15 days.

Natural and Artificial Ponds

Two ponds (Each one with and without *Gambusia*) with *Hydrilla* were being studied to evaluate the role of the Gambusia fish on mosquito larvae control. Water samples were collected from these ponds and used for analysis of microbial count.

Two fresh ponds were created without *Hydrilla*. Bacterial and Fungal counts were adjusted to 8×10^{10} CFU/ml each using probiotic *B.subtilis* and *A.niger* spores. Water samples were collected at every alternative day and used for microbial count.

Microbial Control : Laboratory Experiments

To check the microbe specific feeding ability of *G.holbrooki*, 2 types of controlled laboratory experiments were carried out.

In the initial experiment 300 ml of pond water was taken in a 500 ml beaker to which 1 ml nutrient broth and 2-3 small *Hydrilla* branches were added. To this selected pathogenic (*E.coli, P.aurogenosa,* and *S.aureus*) and non pathogenic (*B.subtilis* and *S.cerevisiae*) microorganisms were inoculated in order to get the microbial population of 8×10^{10} CFU/ml. In this beaker 15 healthy *Gambusia* fishes (size 2±0.5 cm) were taken and the beakers were closed with a thin cloth.

In another experimental set up, in 500 ml beaker 300 ml of pond water containing specific microorganisms and 15 *Gambusia* fishes were taken. *Hydrilla* and broth media were not taken and also not closed at the top with cloth. Water samples were collected every alternative day up to a month and specific microorganisms load was determined by microbiological methods.

Microbiological Analysis of Water

One ml of water sample was serially diluted with sterile distilled water and the spread plate technique was used to make triplicate aerobic plate count (APC), spore count (SC), Entero bacteriaceae count (EC) and coliform count (Macconkeys agar) using standard microbiological methods (Table 2.1).

For water samples of beakers with specific microbes only the count of selected microorganism was carried out using selective media/selective method as mentioned in Table 2.1. Colony counts with typical characteristics on selective media were used for determination of particular microorganisms load and other colonies were ignored.

Table 2.1: Microbiological Methods

Count Type	*Medium*	*Incubation Temp*	*Incubation Time*
Aerobic–bacteria	Nutrient agar	37°C	48
Aerobic–Fungi	Potato dextrose agar	30°C	72
Enterobacteriaceae count EC	Violet red bile glucose agar	37°C	24
Lactose fermentors	Macconkeys agar	37°C	24
Spore count	Spore count plate agar	37°C	48
E.coli	EMB Agar	37°C	24
Pseudomonas	NA with cetramide	37°C	48
Staphylococcus	Mannitol agar	37°C	24
Bacillus	Preheated water on NA	37°C	24
S. cerevisiae	YEPD with Ampicillin pH5	30°C	48

Characterization of Pond Water Microbes

Based on the colony morphology of microbes, 25 bacterial and 6 fungi were isolated from natural ponds and sub cultured on suitable agar media. The bacterial isolates were characterized by colony morphology, microscopy and biochemical characteristics as explained by Fischer *et al.* (1986). Fungi were characterized by growth morphology, microscopy and sugar fermentation patterns.

Results

Microbial Load in Natural Pond

In the natural pond, which is under study for mosquito control using *G. holbrooki*, microbial count is significantly less than the pond under study without *Gambusia*. Spore count and coliform counts were minimum in *Gambusia* pond over the control (Table 2.2). In the characterization process *Staphylococcus, E. coli* and *Pseudomonas* were not detected in *Gambusia* pond where as they are in huge number in control pond.

Table 2.2: Microbial Prevalence of Natural Pond

Sample	*APC Bacteria*	*APC Fungi*	*SC*	*EC*	*CC*
Natural Pond with Gambusia	4×10^5	12	220	5×10^4	3
Natural Pond without Gambusia	2×10^{10}	5×10^3	6×10^6	7×10^3	7×10^3

APC: Aerobic plate count; ANC: Anaerobic plate count; SC: Spore count; EC: *Enterobacteriaceae* count; CC: Coliform count.

Microbial Control in Artificial Ponds Using *Gambusia*

To check the Bacteria and Fungi population reduction at different time intervals, *Gambusia* fish were taken in a pond without *Hydrilla* and with microbes (Bacteria and Fungi each at 8×10^{10} CFU/

ml). A control was also maintained without *Gambusia*. In both the ponds viable count of microorganisms was decreased with time but decrease in *Gambusia* ponds was very rapid. In *Gambusia* pond maximum reduction of microbes was noted with in 20 days where as in control pond only 10 per cent reduction of bacteria and no reduction in fungi was noted (Table 2.3).

Table 2.3: Microbial Control in Artificial ponds

Day	Gambusia Pond		Control Pond	
	Bacteria	Fungi	Bacteria	Fungi
0	8×10^{10}	8×10^{10}	8×10^{10}	8×10^{10}
2	2×10^{10}	2×10^{8}	5×10^{10}	2×10^{11}
4	5×10^{9}	5×10^{6}	7×10^{10}	9×10^{13}
6	1×10^{9}	3×10^{5}	8×10^{10}	8×10^{13}
8	7×10^{8}	1×10^{5}	9×10^{10}	9×10^{13}
10	1×10^{9}	5×10^{4}	6×10^{10}	7×10^{14}
12	8×10^{8}	1×10^{4}	3×10^{10}	9×10^{14}
14	2×10^{8}	7×10^{3}	8×10^{9}	8×10^{14}
16	8×10^{7}	1×10^{3}	6×10^{9}	6×10^{13}
18	5×10^{6}	3×10^{2}	5×10^{9}	5×10^{13}
20	2×10^{5}	11	2×10^{9}	2×10^{13}
22	8×10^{4}	2	8×10^{8}	8×10^{12}
24	2×10^{4}	0	5×10^{8}	5×10^{11}
26	5×10^{3}	0	8×10^{7}	8×10^{11}
28	4×10^{3}	0	7×10^{7}	3×10^{11}
30	5×10^{3}	0	7×10^{7}	8×10^{10}

Microbial Control in Laboratory Condition

Initially natural pond type of condition are created in beaker with *Hydrilla;* in which 1 ml nutrient media was taken and inoculated with known microorganisms; in order to know anti-microbial activity of *Gambusia* on selected microbes. From the tested organisms almost complete (>99 per cent) reduction of *E.coli, B.subtillis* and yeast was noted where as *P. aurogenosa,* and *S. aureus* are reduced significantly (99 per cent). In Control beakers reduction in microbial population is minimal (10 per cent) (Table 2.4).

In other laboratory experiments; media and *Hydrilla* were not taken and kept open similar to a natural water body in villages. Here the results are same like the above but severity of controlling of microbes is more. Hence less time was taken to reduce the initial population to less than 0.01 per cent (Table 2.5).

Discussion

G.holbrooki is planktivorous fish with an ability to control microorganisms in water (Angler *et al.*, 2002). Its natural abundance, easy growth, simplicity of capture and suitable for tests under various conditions makes it suitable for experimentation. Many countries are using *Gambusia* to manage malaria (Rozendaal, 1997). A self maintenance population of Gambusia in village ponds is convenient (Ghosh and Dash 2007). As *Gambusia* is widely being used for mosquito larvae control and it is

Table 2.4: Microbial control of *Gambusia* Fish in Artificial Conditions (Beakers with *Hydrilla*)

Day	*E. coli*		*B. subtilis*		*P. aurogenosa*		*S. aureus*		*S. cerevisiae*	
	Gambusia	*Control*	*Gambusia*	*Control*	*Gambusia*	*Control*	*Gambusia*	*Control*	*Gambusia*	*Control*
0	8×10^{10}	8×10^{10}	8×10^{10}	8×10^{10}	8×10^{10}	8×10^{10}	8×10^{10}	8×10^{10}	8×10^{10}	8×10^{10}
2	7×10^{13}	5×10^{13}	5×10^{10}	4×10^{11}	7×10^{10}	8×10^{10}	9×10^{8}	5×10^{8}	5×10^{9}	2×10^{11}
4	5×10^{8}	6×10^{12}	3×10^{8}	7×10^{11}	5×10^{10}	2×10^{11}	2×10^{8}	6×10^{9}	8×10^{9}	4×10^{11}
6	8×10^{7}	8×10^{11}	5×10^{6}	3×10^{10}	1×10^{10}	3×10^{11}	4×10^{8}	9×10^{8}	6×10^{8}	9×10^{10}
8	5×10^{5}	9×10^{10}	4×10^{5}	5×10^{8}	7×10^{9}	2×10^{11}	7×10^{7}	2×10^{9}	5×10^{8}	5×10^{10}
10	2×10^{6}	5×10^{10}	2×10^{4}	8×10^{7}	2×10^{9}	5×10^{11}	2×10^{8}	8×10^{9}	2×10^{8}	7×10^{10}
12	3×10^{4}	3×10^{10}	5×10^{3}	5×10^{7}	3×10^{8}	8×10^{11}	5×10^{8}	6×10^{8}	6×10^{7}	2×10^{10}
14	8×10^{3}	8×10^{9}	8×10^{2}	7×10^{8}	1×10^{8}	3×10^{10}	8×10^{7}	5×10^{8}	5×10^{5}	8×10^{9}
16	2×10^{3}	6×10^{9}	200	4×10^{7}	3×10^{7}	4×10^{9}	3×10^{7}	8×10^{7}	2×10^{5}	3×10^{9}
18	4×10^{2}	5×10^{9}	52	3×10^{7}	5×10^{7}	2×10^{8}	8×10^{6}	5×10^{7}	3×10^{4}	8×10^{8}
20	15	2×10^{9}	2	8×10^{6}	7×10^{6}	3×10^{8}	2×10^{6}	2×10^{7}	2×10^{2}	2×10^{7}
22	18	8×10^{8}	0	2×10^{7}	8×10^{6}	8×10^{7}	2×10^{6}	8×10^{6}	54	8×10^{6}
24	21	5×10^{7}	0	5×10^{6}	5×10^{6}	7×10^{7}	3×10^{6}	1×10^{7}	2	5×10^{6}
26	25	5×10^{7}	0	3×10^{6}	5×10^{6}	5×10^{7}	2×10^{6}	5×10^{6}	0	8×10^{5}
28	24	4×10^{7}	0	5×10^{6}	4×10^{6}	4×10^{7}	1×10^{6}	2×10^{6}	2	7×10^{5}
30	17	5×10^{7}	0	8×10^{5}	5×10^{6}	8×10^{6}	2×10^{6}	3×10^{6}	1	5×10^{5}

Table 2.5: Microbial Control of *Gambusia* Fish in Artificial Conditions (Beakers without *Hydrilla*)

Day	*E. coli*		*B. subtilis*		*P. aurogenosa*		*S. aureus*		*S. cerevisiae*	
	Gambusia	*Control*	*Gambusia*	*Control*	*Gambusia*	*Control*	*Gambusia*	*Control*	*Gambusia*	*Control*
0	8×10^{10}	8×10^{10}	8×10^{10}	8×10^{10}	8×10^{10}	8×10^{10}	8×10^{10}	8×10^{10}	8×10^{10}	8×10^{10}
2	9×10^{9}	5×10^{10}	7×10^{9}	4×10^{10}	6×10^{10}	8×10^{10}	7×10^{9}	2×10^{10}	3×10^{9}	2×10^{10}
4	4×10^{8}	8×10^{9}	3×10^{8}	9×10^{9}	4×10^{10}	7×10^{10}	6×10^{8}	8×10^{9}	1×10^{9}	8×10^{9}
6	7×10^{7}	5×10^{9}	4×10^{7}	4×10^{9}	9×10^{9}	2×10^{10}	7×10^{7}	2×10^{9}	5×10^{8}	1×10^{9}
8	3×10^{6}	7×10^{8}	3×10^{5}	7×10^{8}	7×10^{9}	8×10^{9}	1×10^{8}	7×10^{8}	3×10^{7}	7×10^{8}
10	7×10^{5}	2×10^{8}	2×10^{4}	5×10^{8}	2×10^{9}	5×10^{9}	8×10^{7}	4×10^{8}	9×10^{6}	5×10^{8}
12	5×10^{4}	5×10^{7}	3×10^{3}	8×10^{7}	8×10^{8}	1×10^{9}	6×10^{7}	1×10^{8}	2×10^{6}	7×10^{7}
14	9×10^{3}	2×10^{7}	6×10^{2}	5×10^{7}	9×10^{7}	7×10^{8}	2×10^{7}	9×10^{7}	5×10^{4}	3×10^{7}
16	8×10^{2}	8×10^{6}	100	3×10^{7}	5×10^{7}	4×10^{8}	5×10^{6}	5×10^{7}	8×10^{3}	9×10^{6}
18	2×10^{2}	7×10^{6}	42	4×10^{7}	3×10^{7}	2×10^{8}	2×10^{6}	3×10^{7}	4×10^{2}	8×10^{5}
20	75	5×10^{6}	15	9×10^{6}	8×10^{6}	8×10^{7}	5×10^{5}	6×10^{6}	5×10^{1}	5×10^{4}
22	9	3×10^{6}	3	5×10^{6}	4×10^{6}	3×10^{7}	3×10^{5}	3×10^{6}	34	2×10^{3}
24	2	3×10^{6}	0	3×10^{6}	3×10^{5}	9×10^{6}	1×10^{5}	1×10^{6}	6	8×10^{2}
26	1	1×10^{6}	0	2×10^{6}	8×10^{4}	5×10^{6}	6×10^{4}	7×10^{5}	0	489
28	0	2×10^{6}	0	2×10^{4}	4×10^{4}	6×10^{5}	3×10^{4}	2×10^{5}	0	117
30	0	2×10^{6}	0	8×10^{3}	5×10^{4}	3×10^{5}	2×10^{4}	9×10^{4}	1	90

Figure 2.1

reported to keep microbial population constant and reduce (Adrian *et al.*, 1999), it is advisable to evaluate the usage of *Gambusia* for the control of pathogenic microorganisms in water. Generally water is contaminated with human activities by skin contact and fecal contamination. *P.aurogenosa, S.aureus* species are contaminated by infected skins contact with water and *E.coli* by fecal contamination. Hence in the present study pathogenic *P.aurogenosa, S.aureus, E.coli* along with non-pathogenic *B.subtillis* and *S.cervesae* were used for the microbial control studies. Studies of Angeler *et al.* (2002) showed reduced number of bacteria, planktons and autotrophic pico planktons in semi-arid water bodies with *Gambusia*. But control of specific microbes was not studied. In the present study, we confirm the significant reduction of non-pathogenic *B.subtilis, S.cerevisiae* and pathogenic *E.coli, P.aurogenosa* and *S.aureus*. *Gambusia* also known to control the population of commercial fishes and crustaceans. Release of *Gambusia* in fish crops and prawn cultivating ponds may be fatal. Hence it is suggested to use *Gambusia* fish in potable and cattle usage water bodies only.

To achieve an impact on malaria incidence, breeding of vector mosquitoes should be nearly eliminated as a low-density vector population can sustain transmission (Rozendaal, 1997). Every breeding habitat should be identified and targeted to eliminate mosquito larvae. This requires constant monitoring and vigilance. Insecticide (DTT) use for vector control provided quick relief and the use of fish in vector control programs was overlooked. But now many mosquitoes developed resistance to DTT and toxicity of DTT is well established. Hence many countries are encouraging mosquito control using *Gambusia* (Rozendaal, 1997). In India, larvivorous fish are used as a major component of the integrated malaria control program (Sharma, 1984). As the people are starting the usage of Gambusia and if we make them aware of additional application of microbial control, malaria control and many water borne diseases control is possible at a time.

References

Adrian, R. Schneider and Olt, B., 1999. Top-down effects of crustacean zooplankton on pelagic microorganisms in a mesotrophic lake. *J. Plankton Res.*, 21: 2175–2190.

Angeler, D.G., Rodrigo, M.A., Carrillo, S.S. and Cobelas, M.A., 2002. Effects of hydrologically conffned fishes on bacterioplankton and autotrophic picoplankton in a semiarid marsh *Aquat. Microb. Ecol.*, 29: 307–312.

Boswell, E., Tiwari, S.N. and Ghosh, S.K., 2005. Feasibility of global positioning systems in mapping of mosquito breeding sites for the control of malaria vectors using larvivorous fish in Karnataka State, India. *Trans. R. Soc. Trop. Med. Hyg.*, 99: 944.

Fischer, G., Kunemund, V. and Schachner, M., 1986. Neurite outgrowth patterns in cerebellar microexplant cultures are affected by antibodies to the cell surface glycoprotein LI. *J. Neurosci.*, 6: 605–612.

Ghosh, S.K. and Dash, A.P., 2007. Larvivorous fish against malaria vectors: A new outlook. *Trans. R. Soc. Trop. Med. Hyg.*, 101: 1063–1064.

Nunes, B., Gaio, A.R., Carvalho, F. and Guilhermino, L., 2008. Behavior and biomarkers of oxidative stress in *Gambusia holbrooki* after acute exposure to widely used pharmaceuticals and a detergent. *Ecotoxicol. and Environ. Safety*, 71: 341–354.

Peng, K., Luo, C., Lou, L., Li, X. and Shen, Z., 2008. Bioaccumulation of heavy metals by the aquatic plants *Potamogeton pectinatus* L. and *Potamogeton malaianus* Miq. and their potential use for contamination indicators and in wastewater treatment. *Scine Total Environt.*, 392(1): 22–29.

Rozendaal, J.A., 1997. *Vector Control: Methods for Use by Individuals and Communities.* World Health Organization, Geneva.

Sharma, V.P., 1984. Role of fishes in vector control in India. In: *Larvivorous Fishes of Inland Ecosystems,* (Eds.) V.P. Sharma and A. Ghosh. Malaria Research Centre (ICMR), Delhi, p. 1–19.

Toft, G., Baatrup, E. and Guillette Jr, L.J., 2004. Altered social behavior and sexual characteristics in mosquitofish (*Gambusia holbrooki*) living downstream of a paper mill. *Aquatic Toxicol.*, 70: 213–222.

Van der Grift, B. and Griffioen, J., 2008. Modelling assessment of regional groundwater contamination due to historic smelter emissions of heavy metals. *J. f Conta Hydrol.*, 96: 1–4, 19, 48–68.

Chapter 3

Ultraviolet Radiation Tolerance in Zooplankton Species of Tinquilco Lake (38°S Araucania Region, Chile): Experimental and Field Observations

☆ *Patricio De los Rios, Enrique Hauenstein,*
Patricio Acevedo and Ximena Jaque

Introduction

The zooplankton assemblages in Chilean lakes are characterized by their oligotrophy, due to the native forest and chemical composition of the soil of their basin, that avoid the nutrient entry from the land to the water (Oyarzún *et al.*, 1997; Soto and Campos 1995; Steinhart et *al.*, 1999; 2002). Nevertheless, some lakes located mainly between 38 to 41°S has been observed an oligotrophy to mesotrophy transition due to the replacing of the native forest of their basin by agricultural zones, towns and industries (Soto and Campos 1995; Woelfl *et al.*, 2003; Soto 2002). In spite of these conditions, in southern Chile there are many pristine lakes, located in Andes mountains in zones with limited access, some of these are included in protected areas of Chilean government, and in this condition these lakes are pristine and unpolluted, that would be the primitive condition of southern Chilean lakes (Steinhart *et al.*, 1999; 2002).

Nevertheless, currently in Southern Patagonia, it was observed an increase of penetration of ultraviolet radiation (UVR), due to stratospheric ozone depletion (Marinone *et al.*, 2006; Villafañe *et al.*, 2001). The UVR cause alterations in freshwater ecosystems, because it can penetrate in surface layers in transparent waters, such as lakes with oligotrophic status and with dissolved organic carbon or humic acids concentration (Morris *et al.*, 1995). Both conditions of oligotrophy and low concentration of humic substances are common in southern Argentinean and Chilean lakes (De los Rios and Soto

2006; Modenutti *et al.*, 1998). In this scenario, on the basis of observation for Andean Argentinean lakes, the UVR can penetrate at approximately an average of 10 m depth (Morris *et al.*, 1995). In this scenario of penetration of UVR radiation at surface, it is affected the zooplankton, that develops photoprotective strategies for avoid the UVR damage (Villafañe *et al.*, 2001, Grad *et al.*, 2003). In southern Patagonian lakes, that have zooplanktivorous fishes populations, the zooplankton is characterized by its transparent body as protective strategy for avoid the visual predation by fishes (De los Rios and Soto 2006; Modenutti *et al.*, 1998). For the transparent zooplankton the exposition to UVR can have lethal effects (Villafañe *et al.*, 2001; De los Ríos and Soto 2005; De los Ríos 2004) and for avoid the damage of UVR, these species develop vertical migration as a negative phototaxis against UVR (Villafañe *et al.*, 2001; Storz and Paul 1998; Rhode *et al.*, 2001; Alonso *et al.*, 2004). On an community ecology view point, the zooplankton assemblage would be conditioned by the tolerance to UVR exposure, then for Chilean Araucanian lakes, it was described that in oligotrophic status, calanoids are more tolerant to UVR radiation that daphnids, whereas in mesotrophic status the daphnids have more tolerance to UVR radiation (De los Rios and Soto 2005). These results would explain the dominance of daphnids in southern Chilean lakes with transition to oligotrophy to mesotrophy (De los Rios and Soto 2006; Villalobos 2002).

The present study was done in Tinquilco lake, an oligotrophic depth lake (Zmax < 40 m), located within the Huerquehue National Park, a protected area of Chilean Government. This park is within a mountain zone, with difficult access, that have native forest of *Nothofagus alpina* (Poepp. and Endl.), *N. pumilio* (Poepl. and Endl.) Krasser, and *N. dombeyi* (Mirb) Oerst., between 700 to 1200 m a.s.l *Araucaria araucaria* (K. Koch) is dominant in the forest at 1300 m a.s.l (De los Ríos *et al.*, 2007). This park, has numerous lakes, and only eight of these are accessible by mountain ways. Tinquilco lake (39° 10′ 00″ S; 71° 43′ 25″ W; 763 m a.s.l) is located in the main access of the park (De los Ríos *et al.*, 2007), and it is the exclusive water body with embarkation availability for study pelagic environment. Considering their relative unpolluted status of Tinquilco lake (Steinhart *et al.*, 2002), it was study the effects of exposure to natural ultraviolet radiation on composition of zooplankton community by field observations and a short term experiment.

Material and Methods

Field Observations

The field work was done the 24th and 25th March 2006, it was collected zooplankton samples at the central zone of the Tinquilco lake, it was collected zooplankton with vertical hauls using an Apstein net of 20 cm diameter and 80 µm mesh size. The vertical hauls were done at triplicate from surface to 5 and 20 m depth, and it was estimated the zooplankton abundance in the following transects surface-5 m, and surface-20 m. This procedure was done at 0300 PM and 0600 PM of 24th March, and 0900 AM and 0100 PM of 25th March. The zooplankton specimens were fixed in absolute ethanol, identified with specialized literature (Araya and Zúñiga 1985; Bayly 1992), and counted. Also, it was collected the 24th March, a sample for estimate the humic acids (Kornberg 1999) and chlorophyll (Wetzel and Likens, 1991) concentration at surface, the attenuation coefficient of UVR (Morris *et al.*, 1995), and it was measured the depth of Secchi disk. It was calculated the Shannon diversity index for all samples, and it was compared the pair of Shannon index for both sampled depth in each sampled period, in according to descriptions of Zar (1999).

Experimental Work

It was designed a single experiment considering the species *Daphnia pulex*, *Ceriodaphnia dubia* and *Boeckella gracilis* as main component of zooplankton assemblage of Tinquilco lake. The experiment consisted in nine plastic beakers filled with 500 ml of water of Tinquilco lake, that were inoculated with ten specimens of each species, the experiment was done in triplicate, and began at 1200 hrs of 24th March. It was counted the dead individuals at 0300 PM and 0500 PM. The obtained data were analysed in the following steps: 1) Comparison for each species at two sampled period: this analysis was applied by T student test 2) Comparison for the three species in both sampled period: this analysis was done using an ANOVA, and a Tukey multiple comparison test, when were observed significant differences. Previously were verified the conditions of normality and homocedasticity (Zar 1999). All statistics test were applied using the Software Statistica 5.0.

UVR measurements: due to the access problems of the studied site, it was considered the UVR measured of Temuco (38° 41' S–72°35' W), at 80 km at West of studied site, for the summer 2005-06 (December-March), were done in situ with a spectroradiometer Li-Cor model 1800. The data obtained included an spectral radiation between 300 and 1100 nm. The spectroradiometer is calibrated using an optical portatile calibrator model 1800-02 that can measure the optical radiation between 300 and 1100 nm. This calibrator model has a tungsten calibration halogen hourglass operated at 3150° K, this hourglass is calibrated in according to the indications of the National Bureau of Stantard (NBS) of Unites States of America. It was considered the measurements of UVR in Temuco, as approximation and reference data, in spite of the spatial difference (De los Ríos *et al.*, in press), using similar procedure described in Cabrera *et al.* (1997), who studied Negra Lagoon, and used as reference UVR measurements in Santiago, at 100 km from the studied site.

Results

The results of field observations, first, a low humic acids (0,1mg/l) and chlorophyll (1,5 µg/l) concentration, and Secchi disk depth of 8 m. The results of UVR radiation in Temuco, are exposed in Table 3.1, and indicated high values. The results of zooplankton abundances, denoted high daphnid relative abundance (Table 3.2) that would supported by chlorophyll concentrations.

Table 3.1: UV-B Radiation, Daily Maximum (W/m²) and Dose (kJ/m²), for Temuco (38°41' S; 72°35' W) between December 2005 and March 2006

	UVB (280-320 nm)	
	Dose (J/m²)	*Maximum Irradiance (W/m²)*
28 December 2006	107,6	4,3
29 December 2006	86,8	3,5
03 January 2006	73,8	3,4
10 January 2006	91,4	4,0
25 January 2006	100,8	4,2
26 January 2006	97,6	4,1
01 February 2006	99,4	4,2
13 February 2006	101,9	3,0
02 March 2006	69,5	3,2
03 March 2006	59,4	2,5
17 March 2006	45.3	2.8
24 March 2006	42.6	2.1
27 March 2006	51.5	2.4

Table 3.2: Abundance (in ind./l.) of Crustacean Zooplankton Species, and Shannon Biodiversity Index for Two Sampling Depth Obtained in Tinquilco Lake during the Studied Periods

	24th March 2006			
	0300 PM		*0600 PM*	
	0-5 m Depth	*0-20 m Depth*	*0-5 m Depth*	*0-20 m Depth*
Ceriodaphnia dubia	5.032	14.268	15.898	7,669
Daphnia pulex	0.764	0.679	0.815	0,178
Boeckella gracilis	0.000	1.755	0.000	1,223
Neobosmina chilensis	0.127	0.057	0.082	0.000
Mesocyclops longisetus	0.000	0.000	0.001	0,076
Shanon Index	0.737	1.211	1,219	0,936
	T obs: 6.443 > T table: 1.978		T obs: 5,398 > T table: 2.014	
	25th March 2006			
	0900 AM		*0100 PM*	
	0-5 m Depth	*0-20 m Depth*	*0-5 m Depth*	*0-20 m Depth*
Ceriodaphnia dubia	1.975	15.334	1.975	15,334
Daphnia pulex	2.357	1.433	2.357	1,433
Boeckella gracilis	0.000	1.218	0.000	1,218
Neobosmina chilensis	0.000	0.072	0.000	0,072
Mesocyclops longisetus	0.000	0.000	0.000	0.000
Shanon Index	0.034	1,215	0.567	1.244
	T obs: 6.013 > T table: 1.960		T obs: 9.169 < T table: 1.983	

Also, it denoted low species absolute abundance between surface to 5 m depth during practically all sampled period, with exception to 0600 PM, period that denotes high species abundance in comparison to the other periods (Table 3.2). The analysis of Shannon index, first, revealed a low species richness (Table 3.2), and significant differences between species richness at two sampling depths. The results of experimental evidence denoted the existence of significant differences in absolute at three ("p" < 0,03; Table 3.3) and five hours ("p" < 0,03; Table 3.3), whereas if we compared the mortality results for each species it was significant differences only for *C. dubia* spp (p = 0,024; Table 3.3), whereas had significant differences for *B. gracilis* (p < 0.001; Table 3.3) and *Daphnia* spp (p = 0.374; Table 3.3). The results of Tukey multiple comparison test denoted significant differences at three hours between *B. gracilis* and *Daphnia* spp. (p = 0.031; Table 3.3). Whereas at six hours *Daphnia* spp., had significant differences with *B. gracilis* (p = 0.010; Table 3.3) and *C. dubia* (p < 0.010; Table 3.3).

Discussion

The results of field observations would agree with literature descriptions that indicated the high penetration of ultraviolet radiation in surface layers, and this condition generate a negative phototaxis response of zooplankton species, that migrate to depth zones without exposition to UVR radiation (Leech and Williamson 2001, Leech *et al.*, 2005, Rhode *et al.*, 2001, Villafañe *et al.*, 2001, Alonso *et al.*,

2003). Although there are not field observations of penetration of UVR in Tinquilco lake, the results the low humic acid concentration, would support that UVR would not absorb in the water column, these results are similar to descriptions for Argentinean Patagonian lakes (Morris *et al.*, 1995). The relative high presence of daphnids in the zooplankton assemblage (Table 3.2) would be explained to the relative mesotrophic status of the studied site (De los Rios *et al.*, 2007). Field and observations for Chilean lakes revealed that daphnids are abundant in mesotrophic status, and deep lakes, because this last condition would provide protection against UVR penetration (De los Ríos and Soto 2006, 2007; Soto and De los Ríos 2006). The presence of *B. gracilis* in depth layers (> 5 m) is a remarkable situation, it is probable that this specie would be more vulnerable to UVR radiation in comparison to *D. pulex* and *C. dubia* (Table 3.3) the literature about UVR tolerance in Patagonian, indicated that calanoids corepods are more tolerant to UVR radiation in comparison to daphnids (De los Rios and Soto 2005, De los Ríos 2005). Nevertheless the results obtained in the present paper do not agree with these literature descriptions, that described a low tolerance to ultraviolet radiation of non pigmented calaniods species such as *B. gracilipes* in comparison to pigmented calanoids species such as *B. gibbosa* and *B. antiqua* (Zagarese *et al.*, 1997a,b). These results would be explained due trophic interactions such as maximum grazing of cladocerans (Martinez, 2000), other option would be carnivorous or omnivorous diet of *B. gracilis* in a similar pattern described for the carnivorous copepods in depth zones for Argentinean Patagonian lakes (Reissig *et al.*, 2004).

Table 3.3: Individual Dead Average (± standard error) for Treatments of Field Experiments Obtained during the Studied Period for the Three Zooplanktonic Species Collected from Tinquilco Lake and Results Statistical Analysis Applied in the Present Study (P values lower than 0.05 denotes significant differences)

	3 hours	*5 hours*	*Results of T Test*
Boeckella gracilis	1.33±0.33	2.33±0.33	p = 0.101
Ceriodaphnia dubia	0.66±0.33	2.33±0.33	p = 0.024
Daphnia spp	0.00±0.00	0.33±0.33	p = 0.374
Results of ANOVA	F = 0.06; p < 0.03	F = 12.00; p < 0.01	

Tukey multiple comparison test for each one of the three species.

	Boeckella gracilis	*Ceriodaphnia dubia*	*Daphnia spp*
Boeckella gracilis	–	1 hour (p = 0.269)	1 hour (p = 0.031)
Ceriodaphnia dubia	6 hours (p = 0.999)	–	1 hour (p = 0.269)
Daphnia spp	6 hours (p = 0.001)	6 hours (p = 0.001)	–

The results of experimental observations, denotes that the calanoid *B. gracilis* would be most vulnerable species to UVR exposure, whereas *Daphnia* spp, was the most tolerant species to UVR exposure. These results would be explained because *B. gracilis* is absent in surface samples for all studied periods (Table 3.2), and probably this species would be more vulnerable to UVR exposure. This result, would agree with observations of De los Rios *et al.* (in press), that indicates the presence of *B. gracilis* in shallow pond, with high chlorophyll "a" and humic acids concentrations, although this species was found hidden in submersed vegetation, that provide protection against UVR exposure (Bursk *et al.*, 2003). Nevertheless, also *C. dubia* was vulnerable to UVR exposure (Table 3.2). If we compared the experiment result with field observations, it was observed a high dominance of *C. dubia* only in samples collected from surface to 20 m depth, whereas not observed a regular pattern in surface samples (Table 3.2). The dominance of *C. dubia* in the more deep samples collected, would agree with the experimental evidence of considerable high mortality of this specie (Table 3.3). The exposed results, revealed the differential tolerance of dominant species in zooplankton assemblages, these results do not agree with literature descriptions that indicated that calanoids copepods are more tolerant to UVR radiation in comparison to daphnids (De los Rios and Soto 2005, De los Rios 2005). These results support the propose described by Marinone *et al.* (2006), that indicated an important role of UVR exposure as regulator of community structure in Patagonian lakes. In according to Marinone *et al.* (2006), in conditions to high UVR exposure, there are low species richness in zooplankton community, because the most vulnerable species would be absent or low abundant (De los Rios and Soto 2005; De los Rios 2005). These descriptions are agreed with field observations that denote a relative low species abundant in surface samples (Table 3.2). Also, similar results are described for littoral microcrustaceans in Huerquehue National Park waterbodies (De los Rios *et al.*, 2007), that described a direct relation between species richness with chlorophyll and humic acids concentration. The obtained results would conclude: first the existence of differential responses to UVR radiation in zooplanktonic species; second, the UVR radiation would have an important role as regulator in zooplankton community structure.

Acknowledgements

The present study was financed by the Research Directorate of the Catholic University of Temuco, Chile (Project DGI-UCT 2005-4-01), and the Research Directorate of the La Frontera University, Chile (Project DIUFRO N°120614). We express the gratitude to the personal of National Forestal Corporation of National Park Huerquehue for the access facilities for their respective parks, and the staff of Agro-industry laboratory (La Frontera University).

References

Alonso, C., Rocco, V., Barriga, J.P., Battini, M.A. and Zagarese, H., 2004. Surface avoidance by freshwater zooplankton: field evidence on the role of ultraviolet radiation. *Limnol. Ocean.*, 49(1): 225–232.

Araya, J.M. and Zuñiga, L.R., 1985. Manual taxonómico del zooplancton lacustre de Chile. *Boletín Informativo Limnológico, Universidad Austral de Chile*, 8: 110.

Bayly, I.A.E., 1992. Fusion of the genera *Boeckella* and *Pseudoboeckella* (Copepoda) and a revision of their species from South America and sub-Antartic islands. *Rev. Chil. Hist. Nat.*, 65(1): 17–63.

Burks, R.L., Lodge, D.M., Jeppensen E. and Lauridsen, T.L., 2002. Diel horizontal migration of zooplankton: Costs and benefits of inhabiting the littoral. *Freshwat. Biol.*, 47(3): 343–365.

Cabrera, S., López, M. and Tartarotti, B., 1997. Phytoplankton and zooplankton response to ultraviolet radiation in a high altitude Andean lakes: Short-versus long term effects. *J. Plankt. Res.*, 19(11): 1565–1582.

De los Rios, P., Hauenstein, E., Acevedo, P. and Jaque, X., 2007. Littoral crustaceans in mountain lakes of Huerquehue National Park (38°S, Araucania region, Chile). *Crustaceana*, 80(4): 401–410.

De los Ríos, P. and Soto, D., 2007. Crustacean zooplancton richness in Chilean Patagonian lakes. *Crustaceana*, 80(3): 285–296.

De los Ríos, P. and Soto, D., 2006. Structure of the zooplanktonic crustaceous Chilean lacustre assamblages: Role of the trophic status and protection resources. *Crustaceana*, 79(1): 23–32.

De los Ríos, P. and Soto, D., 2005. Survival of two species of crustacean zooplankton under to two chlorophyll concentrations and protection or exposure to natural ultraviolet radiation. *Crustaceana,* 78 (2): 163–169.

De los Ríos, P., 2005. Survival of pigmented freshwater zooplankton exposed to artificial ultraviolet radiation and two levels of dissolved organic carbon. *Pol. J. Ecol.,* 53(1): 113–116.

De los Ríos, P., 2004. Lethal effects of ultraviolet radiation on *Neobosmina chilensis* (Cladocera, Bosminidae) exposed to ultraviolet radiation. *Crustaceana,* 77: 989–996.

Grad, G.B., Burnet, J. and Williamson, C.E., 2003. UV damage and photoreactivation: Timing and age are everything. *Photochem. and Photobiol.,* 78(3): 225–224.

Leech, D.M. and Williamson, C.E., 2001. *In situ* exposure to UV radiation alters the depth distribution of *Daphnia. Limnol. Ocean.,* 46(2): 461–471.

Leech, D.M., Williamson, C.E., Moeller, R.E, and Hargreaves, B.R., 2005. Effects of ultraviolet radiation on the seasonal vertical distribution of zooplankton: a data base analysis. *Arch. Hydrobiol.,* 162(4): 445–464.

Kornberg, L., 1999. Contents of humic substances in freshwaters. In: *Limnology of Humic Waters,* (Eds.) J. Kestitalo and P. Eloranta. Backhuys Publishers, Leiden, The Netherlands, pp. 9–10.

Marinone, M.C., Menu Marque, S., Añón Suárez, D., Diéguez, M. C., Pérez, A.P., De los Ríos, P., Soto, D., and Zagarese, H.E., 2006. UVR Radiation as a potential driving force for zooplankton community structure in Patagonian lakes. *Photochem. Photobiol.,* 82(4): 962–971.

Martinez, G., 2000. Conducta alimentaria de *Daphnia ambigua* Scourfield. 1947. *Moina micrura* Kurz 1874 y *Ceriodaphnia dubia* Richard 1895 (Cladocera) frente a un gradiente de concentración de alimento. *Rev. Chil. Hist. Nat.,* 73(1): 47–54.

Modenutti, B.E., Balseiro, E.G., Queimaliños, C.P., Añón-Suarez, D.A., Diéguez, M. and Albariño, R.J., 1998. Structure and dynamics of food webs in Andean lakes. *Lak. Reserv. Res. Manag.,* 3: 179–186.

Morris, D.P., Zagarese, H.E., Williamson, C.E., Balseiro, E.G., Hargreaves, B.R., Modenutti, B.E., Moeller, R.E. and Queimaliños, C.P., 1995. The attenuation of solar UVR radiation in lakes and the role of dissolved organic carbon. *Limnol. Ocean.,* 40(8): 1381–1391.

Oyarzún C.E., Campos, H. and A.Huber. 1997. Exportación de nutrientes en microcuencas con distinto uso de suelo en el sur de Chile (Lago Rupanco, X región). *Rev. Chil. Hist. Nat.,* 70(3): 507–519.

Rhode, S.C., Pawlowski, M. and Tollrian, R., 2001. The impact of ultraviolet radiation on the vertical distribution of zooplankton of the genus *Daphnia. Nature,* 412: 69–72.

Riessig, M., Modenutti, B., Balseiro, E. and Queimaliños, C., 2004. The role of the predaceous copepod *Parabroteas sarsi* in the pelagic food web of a large depth Andean lake. *Hydrobiologia,* 524: 67–77.

Soto, D. and Campos, H., 1995. Los lagos oligotróficos del bosque templado húmedo del sur de Chile. In: *Ecología del Bosque Chileno,* (Eds.) J. Armesto, M. Khalin and M. Villagrán. Editorial Universitaria, Santiago de Chile, pp. 134–148.

Soto, D., 2002. Oligotrophic patterns in southern Chilean lakes: the relevance of nutrients and mixing depth. *Rev. Chil. Hist. Nat.,* 75(2): 377–393.

Steinhart, G.E., Likens, G. and Soto, D., 1999. Nutrients limitation in Lago Chaiquenes (Parque Nacional Alerce Andino, Chile): Evidence from nutrient enrichment experiments and physiological assays. *Rev. Chil. Hist. Nat.,* 72(4): 559–568.

Steinhart, G.E., Likens, G. and Soto, D., 2002. Physiological indicators of nutrient deficiency in phytoplankton in Southern Chilean lakes. *Hydrobiologia*, 489: 21–27.

Storz, U.C. and Paul, R.J., 1998. Phototaxis in water fleas (*Daphnia magna*) is differently influenced by visible and UV light. *J. Comp. Physiol.*, A183: 709–717.

Villafañe, V.E., Helbling, E.W. and Zagarese, H.E., 2001. Solar ultraviolet radiation and its impacts on aquatic ecosystems in Southern Patagonia. *Ambio*, 30: 112–117.

Woelfl, S., Villalobos, L. and Parra, O., 2003. Trophic parameters and meted validation in a lake Riñihue (North Patagonia, Chile) from 1978 through 1997. *Rev. Chil. Hist. Nat.*, 76(3): 459–474.

Zagarese, H.E., Williamson, C.E., Vail, T.L., Olson, O.G. and Queimaliños, 1997a. Long-term exposure of *Boeckella gibbosa* (Copepoda, Calanoida), to *in situ* levels of solar UV-B radiation. *Freshwat. Biol.*, 37(1): 99–106.

Zagarese, H.E., Feldman, M. and Williamson, C.E., 1997b. UV-B induced damage and photoreactivation in three species of *Boeckella* (Copepoda, Calanoida). *J. Plank. Res.*, 19(3): 357–367.

Zar, J.H., 1999. *Biostatistical Analysis*. Prentice may, New Jersey, U.S.A.

Chapter 4

Phytoplankton Diversity in Ramdara Reservoir Near Tuljapur, Maharashtra

☆ *J.S. Mohite and P.K. Joshi*

Introduction

Plankton includes very small organisms which float on the water surface and drift at the mercy of water currents. Those of plant origin are called phytoplankton, the producers belonging to first trophic level while those of animal origin are the zooplankton which are the primary consumers belonging to second trophic level. Phytoplanktons are ecologically significant as they trap radiant energy of sunlight and convert into chemical energy. Many herbivores, mostly zooplankton, graze upon the phytoplankton thus, passing the stored energy to its subsequent trophic levels. The role of phytoplankton in energy budgets of aquatic systems and their importance in establishing their states is well known. Phytoplanktons also are biological indicators of water quality in pollution studies. To summarize, due to their involvement in cycling of energy and matter in an ecosystem, evaluation of phytoplankton population terms of their diversity, density, biomass, spatial and temporal distribution, periodicity and productivity and population turnover, is vital in management of an ecosystem.

The density of phytoplankton in a water body determines the stocking rate of fishes because they are the chief source of food of many economically important fishes. Phytoplankton, due to its key role in the ecosystem of the environment, is directly related to the fish catch potential of a reservoir. An insight into the distribution, composition and succession of phytoplankton gives valuable clue for determining the fishing grounds, selection of suitable species for stocking and determining the level of utilization of the available food by the existing fish stock.

Materials and Methods

Collection of phytoplanktons was done by using a plankton net with 38 cm diameter of the mouth and a bolting silk No. 20 (173 meshes/inch). An inron tube was firmly tied to the tapering end of the

net and the open end of the phytoplankton collecting tube was covered by a piece of bolting silk, securely tied with cotton thread so that phytoplankton collected through the net could be easily transferred into separate plastic bottles. Initial study for taxonomic identification was carried out on fresh phytoplankton. Subsequently,quantitative estimations were made on plankton preserved in 5 per cent formalin. Samples collected from Ramdara reservoir were easily washed with formalin water. For this purpose a glass funnel and a piece of bolting silk were used. But,washing was rather difficult for samples collected from few sites. It is because these samples contained lot of debris, micro and macro-phytes. By using a wash bottle containing formalin water, washing was carried out. Preliminary identification was made by using Pennak (1978) Tonapi (1980) and Agarwal (1999) as basic references.

Results and Discussion

The phytoplankton species occurred in the reservoir during year 2007-2008 is listed in Table 4.1.

Table 4.1: Phytoplankton Diversity in Ramdara Reservoir

Chlorophyceae: *Ankistrodesumus* sp., *Coelastrum* sp., *Closterium* sp., *Pediastrum* sp., *Scenedesmus* sp., *Staurastrum* sp., *Cosmarium* sp., *Chlorella* sp., *Spriogyra* sp., *Ulothrix* sp., *Volvox* sp., *Oedogonium* sp.

Cyanophyceae: *Anabaena* sp., *Chroococcus* sp., *Spirulina* sp., *Microcystis* sp., *Lyngbya* sp., *Nostoc* sp., *Merismopedia* sp., *Oscillatoria* sp.

Bacillariophyceae: *Cyclotella* sp., *Gyrosigma* sp., *Diatomas* sp., *Cymbella* sp., *Melosira* sp., *Fragillaria* sp., *Tabellaria* sp., *Navicula* sp., *Nitzschia* sp., *Pinnularia* sp., *Synedra* sp.

Euglenophyceae: *Euglena* sp., *Phacus* sp.

During the present investigation, from chlorophyceae, *Cosmarium* sp., *Pediastrum* sp., and *Ulothrix* sp. dominated the reservoir. 8 species of cyanophyceae were identified. *Microcystis* sp. and *Lyngbya* sp. dominated the reservoir. Out of 8 species recorded, 6 species exhibited their presence during December 2007 to January 2008 and 5 in February 2008. 11 species from bacillariophyceae were identified. *Navicula, Synedra* sp. and *Cyclotella* sp. dominated the reservoir. Out of 11 species recorded, 8 species exhibited their presence during October to December. Euglenophyceae was represented by 2 species with dominance of *Euglena* sp. This group attained a highest peak in month of November.

References

Agarwal, S.C., 1999. *Limnology*. APH Publishing House, New Delhi, pp. 132.

Pennak, R.W., 1978. *Freshwater Invertebrates of the United States*, 2nd edn. John Willey Sons, New York, 803 pp.

Tonapi, G.T., 1980. *Freshwater Animals of India: An Ecological Approach*. Oxford and IBH Publishing Co., Bombay, p. 167.

Chapter 5

Incidence of Vibriosis in the Indian Magur (*Clarius batracus* L.) in Saline Water Ponds of Haryana: A New Report from India

☆ *T.P. Dahiya and R.C. Sihag*

ABSTRACT

To determine the level of fish disease and causative organism(s) in newly introduced cat fish, Indian magur (*Clarius batracus* L.) in the saline water ponds, a survey of six fish farms/ponds was carried out around Hisar (Haryana) during 2004-2005. About 2.5 per cent of the sampled individuals of this fish were found to show skin lesions of different magnitude. Such infected fishes were dissected and their skin lesions, gills and kidneys were taken out for bacterial isolation. These bacterial isolates were subjected to physico-chemical tests. The results so achieved were analyzed by a computer based specific PIBwin programme to derive the ID scores of the causative bacteria. The ID scores obtained in this programme were compared with the standard scores. This comparison characterized three species of bacteria *viz. Vibrio anguillarum, V. alginolyticus* and *Aeromonas hydrophila* causing skin lesions in this fish. This is a new report of incidence of vibriosis in Indian magur cultured in the saline water ponds in any part of the country.

Introduction

Due to the presence of less spiny bones, the catfishes (*Clarius batracus, C. garipienus*) are becoming popular fish food in the inland areas of the country over the Indian major carps (*Catla catla, Labeo rohita, Cirrhinus mrigala*), common carp (*Cyprinus carpio*), silver carp (*Hypophthalmicthys molitrix*) etc. The former fishes are biologically adapted in waters with high salinity. In Haryana, less availability of freshwater, rising underground brackish water at many places and better market for catfishes has created conditions to introduce Indian magur in several village ponds and fish farms. In these ponds, domestic animals like buffalo, cow, camels etc. add urine and feacal materials to water increasing its organic matter that allows growth of various types of bacterial and other fish pathogens.

Bacterial infections and diseases are very common in intensive fish culture and probably are of the biggest cause of fish health problems (Bullock *et al.*, 1971; Itami and Kusuda 1984; Hambal 1985; Manavasta 1985). Vibriosis is a systemic disease of primarily marine fishes caused by bacteria of the genus *Vibrio* and is a major cause of their mortality (Bruno *et al.*, 1985; Colwell *et al.*, 1983, 1984; Egidius *et al.*, 1984; Fryer *et al.*, 1972; Hacking and Budd 1971; Lewis 1985; McCarthy *et al.*, 1974; Tanaka 1975; Tison *et al.*, 1982). The disease is characterized by superficial ulcerations on all over the body surface. As the infection progresses, the ulcers deepen up to skeleton; leaving hemorrhages on the skin. The wounds are characterized by swollen reddish boundary with grayish pink coloration at the centre (Ghittino *et al.*, 1972; Holm *et al.*, 1985; Park and Chun 1986). The disease has been reported to inflict serious losses to the fisheries industry (Bruno *et al.*, 1985). Indian magur has become a new bio-resource tool of livelihood for the fish farmers of Haryana and Punjab utilizing underground saline water. A basic understanding of the prevalence of any bacterial disease and its iteology in this fish and the identification of causative bacteria will possibly be useful in finding successful solutions to its cure/control. With this background the present study was undertaken.

Material and Methods

In the present investigations, a survey was carried out from July, 2004 to March, 2005 on six fish farms around Hisar (Haryana, India). On a sampling day, one hundred individuals of Indian magur (*Clarius batracus* L.) were randomly captured with hand net and percentage of diseased fishes in the samples was determined. The sampling was done on monthly intervals. The live/moribund/diseased individuals were brought to the Fish Biotechnology Laboratory of Department of Zoology and Aquaculture, CCS Haryana Agricultural University; Hisar (Haryana, India) for recording their disease symptoms. These fishes were dissected and their skin lesions, gills and kidneys were taken out for bacterial isolation. These organs were homogenized in a sterilized macerating tube. The supernatant was poured on nutrient agar medium and subsequently incubated in a BOD incubator at 28±2ºC for 24 hours. Next day, by picking up single colony, streaking was done for obtaining pure culture of the isolates. To diagnose the disease, the bacterial pathogens were isolated and subjected to physico-chemical tests (Krieg and Holt 1984; Ottaviani *et al.*, 2003). The results of these tests were recorded as +ve for presence/growth and–ve for absence/no growth. These results were analyzed with the help of a standard and most reliable PIBwin computer programme *http: //www.som/soton/ac.uk/staff/tub/pibwins* (Bryant 1995). Based on the latter programme, the probabilities identification scores have been assigned to each bacterium. If the 'Id scores' of bacterial isolates of this study were equal to or more than the 'threshold identification score' in the PIBwin then these were considered to be the individuals of same bacterium species (Bryant 1995, 2004).Thus the disease causing bacteria were identified.

Results and Discussion

The diseased individuals of the Indian magur collected from fish farms near Hisar (Haryana) showed skin lesions of different magnitude. These were:

1. Early stage: This was characterized by white spots and superficial ulcers (Figure 5.1).
2. Mid stage: This was characterized by hemorrhages in fins and skin (Figure 5. 2).
3. Late stage: In this stage, the ulcers deepened up to skeleton. The wounds were characterized by swollen reddish boundary (Figure 5.3).

Figure 5.1: Vibriosis in Indian Magur–Early Stage

Figure 5.2: Vibriosis in Indian Magur–Mid Stage

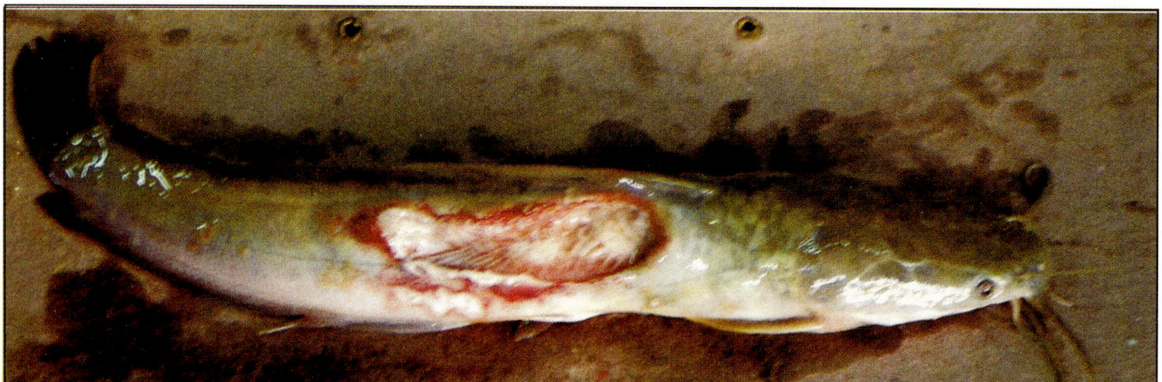

Figure 5.3: Vibriosis in Indian Magur–Late Stage

These symptoms are typical of vibriosis disease caused by bacteria of the genus *Vibrio* and are in consistence with several earlier reports on wide variety of fishes from other part of the globe (Bruno *et al.*, 1985; Colwell *et al.*, 1983, 1984; Egidius *et al.*, 1984; Fryer *et al.*, 1972; Hacking and Budd 1971; Lewis 1985; McCarthy *et al.*, 1974; Tanaka 1975; Tison *et al.*, 1982; Ghittino *et al.*, 1972; Holm *et al.*, 1985; Park and Chun 1986). This contention is further supported by the results of the physico–chemical tests and PIBWin search. Based on the latter tests/search, ID scores of these isolates came out to be 0.741, 0.831

and 0.966, respectively. These ID scores matched the ID scores of three species of bacteria *viz.*– *V. anguillarum, V. alginolyticus* and *A. hydrophila* (Table 5.1). About 2.5 per cent (mean±standard deviation = 2.53±0.44, n = 54) of the sampled fishes were found infected; a considerable loss from the fish farmer's point of view. There is no earlier report on incidence of this disease in India. *V. anguillarum* and *V. alginlyticus* are generally the marine and brackish water bacteria and have been first time detected in the saline water fish farms around Hisar (Haryana, India) causing vibriosis in the Indian magur (*Clarius batracus*). The incidence of vibriosis in the latter fish is a matter of great concern for inland saline water fisheries in India. Incidence of *A. hydrophila* is common in nutrient enriched and polluted fresh as well as brackish waters (Lakshmanaperumalsamy *et al.*, 2005). This is, however, a new report from in land water bodies of the country. This bacterium has been reported to be pathogenic to human being and causes septicemia, wound and ocular, respiratory, bone and intra-abdominal infections (Lakshmanaperumalsamy *et al.*, 2005). Its entry in the saline water fish ponds of present study may be through cattle dung which is a common fertilizer for fish ponds in Haryana.

Table 5.1: Phenotypic and Biochemical Characterization of Pathogenic Bacteria Isolated from Diseased Indian Magur (*Clarius batracus*) in Freshwater Ponds of Haryana during July 2004 to March, 2005

Tests Performed	*Result of Test on Bacterial Isolate Number*		
	1	*2*	*3*
Gram staining	–	–	–
Shape	Curved rod	Curved rod	Rod
Colour of colony	White	White	White
Catalase	+	+	+
Arginine dihydrolase	V	–	+
Vogus proskeur	+	+	+
H/L medium	*	V	V
King's medium	–	V	V
Urease	–	–	–
Simmon's citrate	V	–	–
Nitrate Reduction	+	+	+
Glucose Peptonic Water acid agar	V	+	+
Sorbitol	+	V	+
Mannitol	V	V	+
Sucrose	+	+	+
Starch	V	V	V
Lactose	–	–	–
Maltose	+	+	+
Adonitol	V	+	+
Fructose	+	+	+
Glucose	+	+	+
0 per cent NaCl	–	–	+

Contd...

Table 5.1–Contd...

Tests Performed	Result of Test on Bacterial Isolate Number		
	1	2	3
6 per cent NaCl	+	+	v
Growth at RT	+	+	+
Methyl Red	v	v	+
ID score	0.741	0.831	0.966
Bacteria Identified	*V. anguillarum*	*V. alginolyticus*	*A. hydrophila*

*: Tests not carried out; +: For positive growth; –: Negative for no growth and v: For less growth.

Vibriosis is a systemic bacterial infection of primarily marine and estuarine fishes, caused by bacteria of the genus *Vibrio* (Ross *et al.*, 1968). It is a major cause of mortality in these fishes. Vibriosis is characterized by skin hemorrhages, ulcers or septicemia, anemia and arrhythmia. Outbreaks caused by *V. anguillarum* produce red necrotic or boil like lesions in the musculature, on the fin bases and mouth of Pacific salmonids. Patches also occur on the body surface, and hemorrhages in the gills and the viscera, and the intestinal tract may be inflamed. Serious epizootics caused by *V. anguillarum*, *V. ordalii*, or *V. salmonicida* occur in pink salmon (*Oncorhynchus gorbuscha*)· chum salmon (*O. keta*), Atlantic salmon (*Salmo salar*) (Bruno *et al.*, 1985; Holm *et al.*, 1985), Japanese eel (*Anguilla japonica*) (Tanaka 1975 and Ottaviani *et al.*, 2003), yellowtail (*Seriola quinqueradiata*) (Park and Chun 1986), and ayu (*Plecoglossus altivelis*) (Fryer *et al.*, 1972; Muroga *et al.*, 1984, 1986). *V. anguillarum* caused septicemia in channel catfish (*Ictalurus punctatus*) (Lewis 1985) and vibriosis in marine fishes cultured in Japan (Tanaka 1975; Muroga *et al.*, 1986). *V. anguillarum* was the first species to be identified as a full-fledged fish pathogen, and has now been reported in more than 42 fish species in widely distributed regions (Colwell *et al.*, 1983). Earlier a bacterium isolated from freshwater hatchery was compared with known isolate and was characterized as *V. anguillarum* (Ross *et al.*, 1968). Although *V. anguillarum* is regarded as the dominant species causing vibriosis, several other *Vibrio* species are also pathogenic. *V. ordalii* causes devastating losses among salmon propagated in cage culture in coastal waters of North America's Pacific Northwest (Ross *et al.*, 1968 and Schiewe 1983). Among other species, *V. carchariae* was isolated from a dead sandbar shark (*Carcharhinus plumbeus*), but caused mortality in spiny dogfish (*Squalus acanthias*) 18 h after intraperitoneal injection. *V. alginolyticus* has been found in finfish, shellfish, and marine sediments and has been associated with acute septicemia in sea bream (*Sparus aurata*); *V. damsela* infects damselfish (*Chromis punctipinnis*), a tropical aquarium species, but also infects some species of sharks, and infections have been reported in man (Colwell *et al.*, 1983). *V. vulnificus* is usually encountered as a highly virulent but opportunistic human pathogen, though infects eels and causes development of red patches on the trunk or tail (Tison *et al.*, 1982). *V. salmonicida*, the most recently described species of fish pathogenic vibrios has been shown to be the cause of a septicemia in cultured salmonids in Norway–a disease characterized by a severe anaemia and extensive haemorrhages (Egidius *et al.*, 1984). Dermal lesions accompanied by fin necrosis, haemorrhages, muscle necrosis and focal interstitial and tubular necrosis of the kidneys are common in the winter flounder, *Pseudopleuronectes* americanus (Levin *et al.*, 1972). In diseased rainbow trout (*Salmo gairdneri*), vibriosis results in muscle necrosis, accompanied by interfibrillar haemorrhages, congestion of interfibrillar vessels, and an absence of leucocytic response (McCarthy *et al.*, 1974). In contrast with the generalized septicemia of Pacific salmonids caused by *V. anguillarum*, *V. ordalii* preferentially attacks skeletal and cardiac muscle, the gills, and the gastrointestinal tract. In sharks, *V. carchariae* produces vasculitis in organs of the reticulo-endothelial system and in tropical aquarium species, *V. damsela* characteristically causes skin ulcers. In eels, *V. vulnificus* infections are associated with red patches or swollen lesions on the trunk or tail; in the late stages of infections, histopathologic changes develop in the gills and internal organs (Tison *et al.*, 1982). Salmon infected with *V. salmonicida* show no external pathology but severe anemia develops internally, haemorrhaging occurs in the swim bladder and rectum, and petechiae occur in the caecum and abdominal wall. Although the seven *Vibrio* species reported as fish pathogens may infect many marine and estuarine fishes, cultured fishes are most susceptible because they are often stressed (Colwell *et al.*, 1983, 1984; Egidius *et al.*, 1984; Fryer *et al.*, 1972; Hacking and Budd 1971; Lewis 1985; McCarthy *et al.*, 1974; Tanaka 1975; Tison *et al.*, 1982; Ghittino *et al.*, 1972; Holm *et al.*, 1985).

An outbreak of cold water vibriosis caused 3 per cent mortality in salmon smelts in sea cage at a site in northern Scotland in January 1985. The ambient water temperature was about 7°C with no external sign of disease but a large number of *Vibrio* sp. specimens were isolated from the internal organs of diseased fishes. That was the first report of the disease in Scotland, although it has already been reported in Norway and elsewhere (Bullock *et al.*, 1971). An outbreak of vibriosis with low fish mortality in Italy was reported due to *V. anguillarum* (Ghittino *et al.*, 1972). *Vibrio* strains from diseased turbot fish were isolated at an experiment fish farm on the atlantic coast of northwest Spain and were identified as *V. anguillarum* (Tolmasky *et al.*, 1985; Toranzo *et al.*, 1985). Vibriosis caused severe loss among cultured yellowtail in Korea and *Vibrio* sp. were isolated and identified from kidney (Holm *et al.*, 1985). Although vibriosis is distributed worldwide, *V. ordalii* has thus far been limited to the Pacific Northwest and Japan, and *V. salmonicida* to Norway and Scotland (Holm *et al.*, 1985). In eels, in vibriosis infected individuals, rapidly developing septicemia characteristically occurs; victims have hemorrhages in the fins and striated muscle of the abdominal region coupled with skin ulcerations (Bullock *et al.*, 1971).

The fore going account clearly reveals the dreadedness of the vibriosis disease. Diagnosis of fish vibriosis is based on the isolation and identification of the particular species of *Vibrio* involved. All the *Vibrio* species are gram-negative motile rods and are sensitive to the vibriostat 01129 (2, 4 diamino 6, 7 diisopropylpteridine). Most vibrios ferment glucose anaerogenically. Serological identification tests have been reported for *V. anguillarum*, *V. ordalli*, and *V. salmonicida*. Other species of fish pathogenic vibrios are now differentiated biochemically (Colwell *et al.*, 1983). The present report on the characterization of vibriosis in the Indian magur and identification of its causative bacteria should open doors for its prompt medication and early control, lest the inland fisheries industry in India should not suffer severe economic losses.

Acknowledgements

We are thankful to Dr. Neeru Narula (Senior Scientist, Microbiology) and Dr. S. B. Kalidhar (Professor, Chemistry) for helping us in microbial and chemical tests, and to Dr. S. K. Garg (Head, Zoology and Aquaculture) for providing necessary facilities. Help rendered by Mr. Balraj Sehrawat (Fishery Extension Officer, Haryana) during the field surveys and fish sample collection is gratefully acknowledged.

References

Bruno, W., Hastings, T.S., Ellis, A.E. and Wooten, R., 1985. Outbreak of a cold water vibriosis in Atlantic salmon in Scotland. *Bull. Eur. Assoc. Fish Pathol.*, 5: 62–63.

Bryant, T.N., 1995. Software and identification matrices for probabalistic identification of bacteria (PIB). Available at: *http: //www.staff.med.school.soton.ac.uk/tnb/pib: html.*

Bryant, T.N., 2004. PIBwin: Software for probabalistic identification of bacteria. *J. Appl. Microbiol.*, 97: 1326–1327.

Bullock, G.L., Conroy, D.A. and Snieszko, S.F., 1971. Bacterial diseases of fishes. In: *Diseases of Fishes*, (Eds.) S.F. Snieszko and H.R. Axelrod. TFH Publications Inc Neptune N1, 151 p.

Colwell, R.R. and Grimes, D.J., 1983, 1984. Vibrio diseases of marine fish populations. In: *Diseases of Marine Organisms*, (Eds.) O. Kinne and H. Bulnheim. International Helgoland Symposium, 265–287 pp.

Egidius, E., Soleim, O. and Andersen, K., 1984. Further observations on coldwater vibriosis or Hitra disease (in salmon). *Bull. Eur. Assoc. Fish Pathol.*, 4: 50–51.

Fryer, J.L., Nelson, J.S. and Garrison, R.L., 1972. Vibriosis in fish. *Prog. Fish. Food. Sci.*, 5: 129–133.

Ghittino, P., Andruetto, S. and Vigliani, E., 1972. "Red mouth" enzootic in hatchery rainbow trout caused by *Vibrio anguillarum*. (Enzoozia di "bocca rossa" in trote iridee di allevamento sostenuata da *Vibrio anguillarum*). *Riv. Ital. Piscic. Ittiopathol.* 7: 41–45.

Hacking, M.A. and Budd, J., 1971. Vibrio infections in tropical fish in a freshwater aquarium. *J. Wildl. Dis.*, 7: 273–280.

Hambal, S., 1985. Problems on bacterial diseases and control in Indonesia. In: *Symposium on Practical Measures for Preventing and Controlling Fish Diseases*, July 24–26, Bogor Indonesia.

Holm, K.O., Strom, E., Stensvag K., Raa, J. and Jorgensen, T., 1985. Characteristics of a *Vibrio* sp. associated with the "Hitra disease" of Atlantic salmon in Norwegian fish farms; *Fish. Pathol.*, 20: 125–129.

Itami, T. and Kusuda, R., 1984 Viability and pathogenicity of *Vibrio anguillarum* in sodium chloride solutions of various concentrations isolated from ayu *Plecoglossus altivelis* cultured in freshwater. *J. Shimonoseki. Univ. Fish.*, 32: 33–40.

Krieg, N.R. and Holt, J.G., 1984. In: *Bergey's Manual of Systemic Bacteriology*, 9th edn. Vol. 1. Williams and Williams, Baltimore, London, pp. 1–355.

Lakshmanaperumalsamy, P., Thayumanavan, Th. and Subashkumar, R., 2005. *Aeromonas hydrophila*: A re-emerging pathogen. In: *Marine Microbiology: Facets and Opportunities*, (Ed.) N. Ramaiah. National Institute of Oceanography, Goa, pp. 115–119.

Levin, M.A., Wolke, R.E. and Cabelli, V.J., 1972. *Vibrio anguillarum*: As a cause of disease in winter flounder (*Pseudopleuronectes americanus*). *Can. J. Microbiol.*, 18: 585–1592.

Lewis, D.H., 1985. Vibriosis in channel catfish, *Ictalurus punctatus*. *J. Fish Dis.*, 8: 539–546.

Manavasta, P., 1985. Current fish disease epidemic in Thailand. In: *Symposium on Practical Measures for Preventing Fish Diseases*, July 24–26, Bogor Indonesia.

McCarthy, D.H., Stevenson, J.P. and Roberts, M.S., 1974. Vibriosis in rainbow trout. *J. Wild. Dis.*, 10: 27.

Muroga, K., Lida, M., Matsumoto, H. and Nakai, T., 1986. Detection of *Vibrio anguillarum* from water. *Bull. Jpn. Soc. Sci. Fish.*, 52: 641–648.

Muroga, K., Yamanoi, H., Hironaka, Y., Yamamoto, S., Tatani, M., Jo, Y., Takahashi, S. and Hanada, H., 1984. Detection of *Vibrio anguillarum* from wild fingerlings of ayu, *Plecoglossus altivelis. Bull. Japan. Soc. Sci. Fish.*, 50: 591–596.

Ottaviani, D., Masini, L. and Bacchiocchi, S., 2003. A biochemical protocol for the isolation and identification of current species of *Vibrio* in seafood. *J. Appl. Microbiol.*, 95: 1277–1284.

Park, S.W. and Chun, S.K., 1986. Characteristics of pathogenic *Vibrio* sp. isolated from cultured yellowtail *Seriola quinqueradiata. Bull. Korean. Fish. Soc.*, 19: 47–154.

Ross, A.J., Martin, J.E. and Bressler, V., 1968. *Vibrio anguillarum* from an epizootic in rainbow trout (*Salmo gairdneri*) in the USA; *Bull. Off. Int. Epizoot.*, 69: 1139–1148.

Schiewe, M.H., 1983. *Vibrio ordalii* as a cause of vibriosis in salmonid fish. In: *Bacterial and Viral Diseases of Fish*, (Eds.) J.H. Crosa. Washington Sea Grant Program, Seattle, pp. 31–40.

Tanaka, J., 1975. Vibrio infection of marine fishes. In: *Proceedings of the Third U.S.–Japan Meeting on Aquaculture*, 15–16 October , Tokyo, pp. 113–114.

Tison, D.L., Nishibuchi, M., Greenwood, J.D. and Seidler, R.J., 1982. *Vibrio vulnificus* biogroup 2: New biogroup pathogenic for eels. *Appl. Environ. Microbiol.*, 44: 640–646.

Tolmasky, M.E., Actis, L.A., Toranzo, A.E., Barja, J.L. and Crosa, J.H., 1985. Plasmids mediating iron uptake in *Vibrio anguillarum* strains isolated from turbot in Spain. *J. Gen. Microbiol.*, 131: 1989–1998.

Toranzo, A.E., Barja, J.L. and Devesa, S., 1985. First isolation of *Vibrio anguillarum* biotype I causing an epizootic in reared turbot, *Scophthalmus maximus* in Galicia northwest Spain. *Invest. Pesq.*, 49: 61–66.

Chapter 6

Breakthrough in Breeding and Rearing of Gold Fish (*Carassius auratus*) in Temperate Climatic Conditions of Kashmir Valley

☆ *Sajid Maqsood, Prabjeet Singh and M.H. Samoon*

The Goldfish, *Carassius auratus*, was one of the earliest fish to be domesticated, and is still a jack of all aquarium trades, easily adaptable for both aquarium and open outdoor cement cistern. A relatively small member of the carp family, the goldfish is a domesticated version of a dark-gray/brown carp native to East Asia. It was first domesticated in China (Background information about goldfish. Retrieved on 2006-07-28) and introduced to Europe in the late 17th century. Goldfish can grow to a maximum length of 23 inches (59 cm) and a maximum weight of 9.9 pounds (4.5 kg), although this is rare; few goldfish reach even half this size. The true lifespan of a well-cared goldfish in captivity can extend beyond 10 years (http: //www.livefish.com.au/index.php?main_page=product_info and products_id=647). The goldfish is usually classified as a coldwater fish, and it can live in an unheated aquarium (Pearce, 2006). Temperatures below 10°C (50 °F) are dangerous to goldfish. Conversely, temperatures over 25 °C (77 °F) can be extremely damaging for goldfish (this is the main reason why they shouldn't be kept in tropical tanks) (Smartt, 2001).

Goldfish are popular pond fish, since they are small, inexpensive, colourful and very hardy. In a pond, they may even survive if brief periods of ice form on the surface, as long as there is enough oxygen remaining in the water and the pond does not freeze solid. During winter, gold fish will become sluggish, stop eating, and often stay on the bottom of the pond. This is completely normal; they will become active again in the spring.

Gold fish is successfully bred through out the world (Harrish, 1987). Harvey and Hems (1986) reported that gold fish breeds when they attain an age of one year or even at nine months. These authors have also reported that the gold fish breed up to the age of six to seven years. In tropical countries like India, the gold fish attains maturation in about 5-6 months unlike in temperate regions. Common and comet goldfish can survive and even thrive in any climate in which a pond for them can be created. Introduction of wild goldfish can cause problems for native species (Smartt, 2001). Within three breeding generations, the vast majority of the goldfish spawn will have reverted to their natural olive color. Since they are carp, goldfish are also capable of breeding with certain other species of carp and creating hybrid species (Goldfish. Retrieved on 2006-07-21).

Research by Dr. Yoshiichi Matsui, a professor of fish culture at Kinki University in Japan, suggests that there are subtle differences which demonstrate that while the Crussian carp is the ancestor of the goldfish, they have sufficiently diverged to be considered separate species (Brunner, 2003).

If left in the dark for a period of time, a goldfish will turn almost gray. Goldfish have pigment production in response to light, which is almost like our tanning in the sun. Fish have cells called chromatophores that produce pigments which reflects light, and gives colouration. So, if a goldfish is kept in the dark it will appear lighter in the morning, and over a long period of time will lose its colour.

Live specimens of Gold fish (*Carassius auratus*) measuring 70–80 mm in length were procured from Mumbai (India) in polythene bags filled with 15 liters of water and 10 Kg of oxygen during the month of February 2008. Fishes were acclimatized in glass aquarium tanks (4 × 2.5 × 2.5 feet) containing 100 lit de-chlorinated tap-water. The aquarium tanks were provided with power operated water filter and aerator for maintaining optimum water qualities. Automatic thermostat was also installed in the aquarium to maintain constant water temperature (26-27°C). The fishes were fed on *Tubifex* sp. and *Chironomus* sp. worms and freshwater zooplankton (cladocerans, copepods, etc) twice daily, up to their satiation. After rearing them for a period of 2 month on a protein rich diet in the aquarium tanks, the males and females could be easily distinguished by the presence of tubercles on the pectoral fins and gill cover of male gold fish while the tubercles were absent in the female gold fish. The anal opening in case of male was small, oval shaped and inverted while the female anal opening was larger, more circular in shape and protruding.

The mature male and female in the ratio of 2: 1 were stocked in the glass breeding tanks (3 × 2.5 × 1.5 feet) containing 50 liters of freshwater and artificial fibers or nylon mops were placed in the breeding tanks to act as an egg attaching substratum. In nature, the gold fish spawns usually in the early morning hours. The water temperature was maintained at 25±1°C using an automatic aquarium heater. To stimulate the fish to spawn, water temperature of the breeding tanks was increased by about 2°C. In the breeding tanks, the fish were fed on the formulated diet (35 per cent Crude Protein) and tubifex worms and earthworms thrice a day up to satiation and reared till the brooders were ready to breed.

Before the actual spawning, the colour of mating pairs becomes brighter than before and male begins to chase the female vigorously at random for 3-4 hours bumping into her abdomen. The chasing becomes more intensified with the male getting aggressive and pushing the female gold fish till she releases the eggs. During courtship, the female released the eggs and were fertilized instantaneously by male. The eggs were adhesive and got attached to the nylon mops and some of them sank to the bottom. The spawning took place early in the morning hours. Soon after spawning, the brooders were transferred to other tank as they show tendency to eat their own eggs. The fertilized eggs are about 1.5-2 mm in diameter and are transparent when first laid. The unfertilized eggs were sticky too, but they

Figure 6.1: Male : Female (2 : 1) in the Breeding Tank

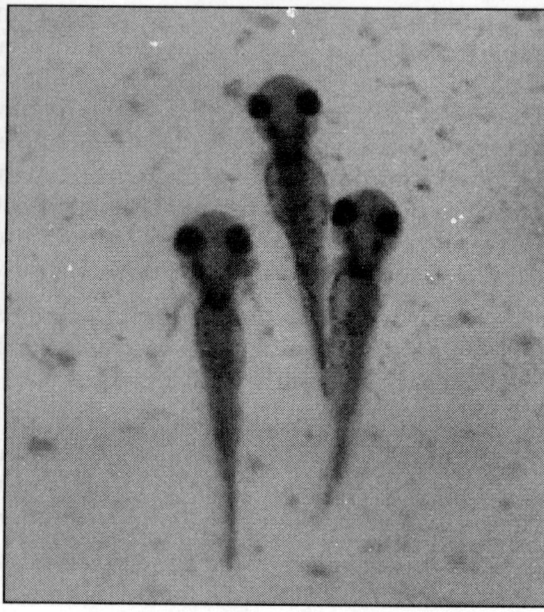

Figure 6.2: 2 Day Old Yolk Sac Fry of Gold Fish

turn opaque, almost milky in few hours and began to decay and get covered by fungus. The unfertilized eggs were regularly siphoned out. The female laid eggs in two batches, about 300-400 eggs in each batch.

Eggs hatched after 6 days at a temperature of 24±1°C. The colour of eggs changed to umber red when they were about to hatch. The embryos hatched out along the region of its back, puling its tail out first, then moving it to release the head out of the eggs shell. At birth, the hatchling was about 4-5mm long with 2 black eyes, a long notochord and a full yolk sac. The yolk sac fry were seen clinging to the substratum and sides of the tank, but some of them were observed resting at the bottom of the tank. The yolk sac of the fry was fully absorbed 4 days after the hatching. After absorption of the yolk sac, air bladder and rudimentary pectoral fin was noticeable. Their air bladder was functional and fry started swimming normally searching for food.

Five-day-old fry measured 7 mm with yolk sac completely absorbed and mouth fully functional. Faint fin rays were visible on caudal, dorsal and ventral fins. For proper development and overall health, an abundance of nutritional food must be fed to the fry right from the start. The fry were fed on the liquefied boiled egg yolk and live food in the form of Daphnai and Moina.

Fry were fed 3-4 times a day for first few weeks of their life. Water quality parameters were monitored regularly and water exchange was carried out after every 2 days. The range of water temperature, pH and hardness of water in the fry rearing tank was 24-26°C, 6.8-7.4 and 170-200mg/lit respectively. Optimum and stable environmental conditions must be maintained to ensure a successful batch of fry. Any sudden change in their environment can easily kill the entire batch. The fry were reared for about 30 days in the same tank supplied with live plankton and liquefied boiled egg yolk. As they grew, the food was gradually shifted to Cladocerans, finely chopped Tubifex worms, powdered food etc. This is the first successful attempt of breeding and rearing of Gold fish in the temperate climatic condition of Kashmir Valley.

References

Background Information About Goldfish. Retrieved on 2006-07-28.

Brunner, Bernd, 2003. *The Ocean at Home*. Princeton Architectural Press, New York.

Goldfish. Retrieved on 2006-07-21.

Goldfish–New World Encyclopedia.

Harrish, J.C., 1987. *Goldfish*. TFH Publications, USA, p. 4–11.

Harvey, F.J. and Hems, 1986. *The Goldfish*. Latimer Trend Co. Ltd., London, p. 196–229.

http: / /www.livefish.com.au/index.php?main_page=product_info and products_id=647.

Lloyd, J. and Mitchinson, J., 2006. *The Book of General Ignorance*. Faber and Faber.

Pearce, Les, 2006. Common gold fish. *Aquarticles*. Retrieved on 2006-06-20.

San Diego Zoo's Got Questions? Animal Group Names. Retrieved on 2007-02-04.

Smartt, Joseph, 2001. *Goldfish Varieties and Genetics: A Handbook for Breeders*. Blackwell Science, 216 pages.

Chapter 7

Effect of Papain on Growth Rate and Feed Conversion Ratio in Fingerlings of *Cyprinus carpio* Under Temperate Climatic Conditions of Kashmir Valley

☆ ***Prabjeet Singh, M.H. Balkhi, Sajid Maqsood and M.H. Samoon***

Introduction

With the present production of over 6.2(2003-04) million tonnes of finfish and shellfish from capture fisheries and aquaculture, Indian fisheries has made a long leap of over eight fold increase in last five decades, not only contributing immensely to the animal protein deficit of country but also playing an important role in global trade. The present status of aquaculture industry poses challenges and opportunities for improvement of production to meet the growing demand in coming years. This demand may be fulfilled by increasing the quality of aqua feeds through improving nutrient content as well as the digestibility of low quality feed by the use of efficient feed additives. Nowadays, feed manufacturers are searching for cheap cheaper alternative protein sources such as plant proteins. Though many of plant origin ingredients have demerits on account of presence of antinutritional factors which have an adverse impact on the digestion and nutrient utilisation of feed however, certain enzymes provide an additional powerful tool that can inactivate antinutritional factors and enhance the nutritive value of plant based protein in feeds. Endogenous enzymes found in the digestive system of fish help to break down large organic molecules like starch, cellulose and protein into simpler substances.

Application of proteolytic enzymes to fish food has stepped into a new era of research to aqua culturists. Herbal based enzyme Papain is a proteolytic enzyme from the cysteine proteinase family. It is derived from papaya leaf, unripe fruit and papaya latex, a milky fluid that oozes out of green papaya. The greener the fruit, more active is the Papain. Papaya latex contains at least four proteolytically active enzymatic components including Papain, chymopapain–A and B and papaya peptidase–A (Glazer and Smith, 1971). Papain enzyme from the unripe fruit act to digest the protein by breaking them down into smaller absorbable components. Aside from its digestive capabilities, this enzyme also possesses antihaemolytic and immuno-stimulant properties (Mowrey, 1986; Kirschman and Dunne, 1984). Papain is the principal and most active enzyme in mature green papaya and possesses a very powerful digestive action superior to pepsin and pancreatin (Ray, 1990).

This enzyme breaks down feed protein molecules into amino acids and promotes growth of fish by increasing availability of aminoacids. It has often been called vegetable pepsin because it contains enzymes similar to pepsin. Papain possesses a very powerful digestive action superior to pepsin and pancreatin. It is one of the most powerful plant proteolytic enzymes that impartially act in protein digestion in acid, alkaline or neutral medium. Papain is allowed to act upon many kinds of proteins and it has a singular distinctive power of converting a portion of protein mass into arginine. Arginine in its natural form has been found to raise this study was carried out with the aim to assess the growth promoting effect of Papain in *Cyprinus carpio* under temperate climatic conditions of Kashmir valley.

Materials and Methods

Collection and Acclimatization of Experimental Fishes

300 uniform sized fingerlings of same age group (10±2 g, 9±2 cm) were procured from the fish farm of Faculty of Fisheries. Fingerlings were acclimatized in FRP rearing troughs (1.8m × 1.8m × 0.5m) at the hatchery complex of the fish farm.

Experimental Diets

Experimental feeds were prepared from the locally available ingredients. Prepared feed was mixed with Papain before feeding. Three levels of Papain (1, 2 and 4 per cent) were used in the present study and fed to three experimental groups after incorporating in the pelleted diet. One control group was maintained on pelleted diet without Papain supplementation. Following was the ingredient composition of the feed:

Ingredients	*Inclusion rate*
Ground nut oil cake	31 per cent
Rice bran	26.23 per cent
Soybean meal	15.9 per cent
Fish meal	4.95 per cent
Wheat flour	19.92 per cent
Vitamin and Mineral mixture	2 per cent

The above formulation contains 29 per cent Crude protein which lies in the range of optimum protein content requirement of the Common carp *i.e.* 25-30 per cent.

Experimental Setup

Twelve rearing troughs of equal size (1.8 × 1.8 × 0.5m) were used and stocked with 25 fingerlings each. Feeding was done @ 3 per cent body weight daily in each experimental feeding group after ascertaining the total group weight on weekly basis. The experiment was conducted for a period of 70 days.

Feeding Combinations

T_1: Feed containing 1 per cent enzyme supplementation.

T_2: Feed containing 2 per cent enzyme supplementation.
T_3: Feed containing 4 per cent enzyme supplementation.
T_4: Control feed without any enzyme supplementation.

Following indices were used to check effect of Papain on Growth and Feed conversion ratio.

$$\textit{Growth rate (g/day)} = \frac{W_1 - W_0}{T}$$

W_1: Fish weight at the end of study
W_0: Fish weight at start of study
T: Time interval in days

$$\textit{Feed conversion ratio (FCR)} = \frac{\text{Weight of feed given}}{\text{Increase in weight of fish}}$$

Statistical Analysis

Experimental data was subjected to the statistical analysis following the CRD and the variation among the treatment means was tested for the significance by analysis of variance (ANOVA) techniques as described by Gomez (1984). Level of significance used for F–test and t–test was P = 0.05 from the table given by Fisher. The critical difference between treatment means, week means and the interaction between week and treatment means have been worked out.

Results

Growth Rate (g/day)

Different levels of enzyme supplementation revealed significant differences among various treatment means on growth rate. The highest growth rate (0.36 g/day) was found with T_2 (feed + 2 per cent Papain) during the 8th week which was significantly different from T_1, T_3 and T_4 ($P<0.05$). Growth rate showed an increasing trend from 2nd to 8th week but decreased during the 10th week and this was true for all the treatments. In T_1 the growth rate showed an increase upto 8th week, the minimum (0.16g/day) during the 2nd week and maximum (0.23g/day) at the end of 8th week and decreased during the 10th week. In general growth rate showed an increasing trend in all the feeding groups but T_1, T_2 and T_3 showed better results as compared to control. The maximum growth rate was recorded (0.36 g/day) during the 8th week thus proving to be the best treatment.

The sum of Means of Treatments and weeks reveal that maximum growth rate (0.29 g/day) was recoded in T_2 and maximum was found during the 8th week (0.26 g/day).

Thus treatment 2nd *i.e.* T_2 showed the best results and the results obtained were significantly different from all other treatments ($P < 0.05$). All the above results are summarised in Table 7.1.

Feed Conversion Ratio (g/g)

Perusal of data (Table 7.2) on FCR revealed significant differences among the treatments on FCR. Lowest FCR (1.6 g/g) was found in T2 (feed + 2 per cent Papain) during the 8th week and it was significantly different from all other treatments ($P < 0.05$). FCR has shown a decreasing trend right from the beginning of 2nd week up to the end of 8th week. Although, better results were also achieved with T_1 and T_3 but T_2 showed the best results. All the three treatments *i.e.* T_1, T_2 and T_3 showed better results as compared to control treatment (T_4 with no enzyme supplementation).

Table 7.1: Effect of Papain Supplementation on Growth Rate (g/day)

Treatment	*2nd week*	*4th Week*	*6th Week*	*8th Week*	*10th Week*	*Mean*
T_1	0.16	0.19	0.22	0.23	0.21	0.20413
T_2	0.21	0.24	0.34	0.36	0.30	0.29167
T_3	0.18	0.18	0.23	0.24	0.18	0. 20080
T_4	0.13	0.18	0.21	0.21	0.15	0.18007
Mean	0.170	0.201	0.251	0.263	0.210	0.219

CD ($P \leq 0.05$); Weeks = 0.014; Treatments = 0.012; Weeks * treatments = 0.028.

T_1: Feed supplemented with 1 per cent Papain; T_2: Feed supplemented with 2 per cent Papain; T_3: Feed supplemented with 4 per cent Papain; T_4: Control feed with no Papain supplementation.

Table 7.2: Effect of Papain Supplementation on Feed Conversion Ratio (g/g)

Treatment	*2nd week*	*4th Week*	*6th Week*	*8th Week*	*10th Week*	*Mean*
T_1	2.7	2.5	2.3	2.1	2.5	2.43
T_2	2.5	2.1	1.9	1.6	2.2	2.07
T_3	2.8	2.6	2.4	2.0	2.5	2.48
T_4	3.2	2.8	2.6	2.4	2.7	2.75
Mean	2.82	2.51	2.31	2.05	2.48	2.43

CD ($P \leq 0.05$); Weeks = 0.014; Treatments = 0.012; Weeks * treatments = 0.028.

T_1: Feed supplemented with 1 per cent Papain; T_2: Feed supplemented with 2 per cent Papain; T_3: Feed supplemented with 4 per cent Papain; T_4: Control feed with no Papain supplementation.

Discussion

The Indian aquaculture will be facing a lot of challenges in the coming decades. The per capita consumption of fish is expected to be doubled by 2020. To meet this growing demand, there is a need of aquaculture production augmentation which can be achieved by the use of herbal growth promoters in a natural way by reducing operational expenditure.

The incorporation of plant based digestive enzyme as an additive in aqua feeds improves the nutrient content as well as the digestibility of low quality feeds in a cost-effective manner. 'Papain' derived from the latex, unripe fruit and leaf of papaya is an appropriate plant based digestive protease. The Papain concentration is maximum in unripe fruit (Watt and Breyer-Brandwijk, 1962) followed by leaf and latex of papaya. Papain is a protein cleaving enzyme and aids in growth and digestion (Ray, 1990).

Feed Conversion Ratio (g/g)

Lowest FCR (1.6 g/g) was recorded in T_2 feeding group which was better than control treatment in which there was no enzyme supplementation (Table 7.2). A comparison among the treatments reveals that T_2 proved to be the best treatment in achieving lowest FCR value. FCR showed an improvement up to 8th week and slightly increased during the 10th week which can be ascribed to certain decrease in temperature. Temperature along with photoperiod appears to be the most important factor, controlling growth and metabolism (Peter and Crim, 1979).

Exogenous application of enzyme resulted in improvement in FCR when compared to control treatment. The reason could be fast metabolism which in turn resulted in better FCR. The present findings are in agreement with those of Rodehutscord and Pfeffer (1995) who found better FCR values by the exogenous application Phytase. Forester *et al.* (1999) found that non inclusion of Phytase doesn't show any improvement in the FCR value. T_1 and T_3 showed lowest FCR value of 2.1 and 2.0 respectively where Papain was supplemented @ 1 per cent and 4 per cent level. This suggests that inclusion of Papain @ 2 per cent gave the best performance.

Lower FCR indicates better utilization of feed by fish. Feed conversion ratio in the present study was reported in the range of 3.2-1.6 which is in agreement with those recorded by Desilva and Anderson (1995).

Growth Rate

As evident from Table 7.1 highest growth rate was found with feed supplemented with Papain @ 2 per cent. In all the feeding groups better growth rate was found in treatments receiving Papain as compared to control treatment in which feed was given without any enzyme supplementation. The reduced growth rate in control treatment could be due to the presence of antinutritional factors in feed (Kakade *et al.*, 1973) which in turn have an adverse impact on growth performance and availability of various dietary compounds (Spinelli *et al.*, 1983, Richardson *et al.*, 1985, Satoh *et al.*, 1989). This suggests that Papain supplementation in the diet may be effective in reducing either antinutritional factors or adverse consequences of phytate from plant origin ingredients of feed which is supported by the finding of Liu (1997). The reduction of phytate-protein complexes in the gut increased nutrient availability could be another explanation for this observation (Liebert and Portz, 2005).

From the results of present investigation, it can be concluded that the dietary inclusion of papain @ 2 per cent in the diet of common carp results in better growth rate and feed conversion ratio. Use of exogenous enzyme in the form of papain in the common carp culture practices in Kashmir valley where the carp farming has started taken pace during last decade can prove very useful in term of growth increment and enhanced fish production. The use of papain can be used in carp farming on regular basis as it has no adverse effect on fish environment.

References

De Silva, S.S. and Anderson, T.A., 1995. *Fish Nutrition in Aquaculture.* Chapmann and Hall Aquaculture Series, 319 pp.

Forester, I., Higgs, D.A., Dosarijh, B.S., Rowashandeli, M. and Parr, J., 1999. Potential for dietary phytase to improve nutritive value of canola protein concentrate and decrease phosphorous output in Rainbow trout (*Onchorhynchus mykiss*) held in 111 °C freshwater. *Aquaculture*, 179: 109–125.

Glazer, A.N. and Smith, E.L., 1971. Papain and other plant sulfhydryl proteolytic enzymes. In: *The Enzymes*, 3rd edn, (Ed.) P.D. Boyer. Academic Press, London, 3: 501–546.

Gomez, K.A. and Gomez, A.A., 1984. *Statistical Procedure for Agricultural Research*, 2nd edn. John Wiley and Sons Inc., New York.

Kakade, M.L., Hoffa, D.E. and Liener, I.E., 1973. Contribution of trypsin inhibitors to the deleterious effects of unheated soybeans fed to rats. *J. Nutr.*, 103: 1172.

Kirschman, J.D. and Dunne, L.J., 1984. Bromelain–Papain. *Nutr. Almanac*, 2: 171.

Liebert, F. and Leandro, Portz, 2005. Nutrient utilization of Nile Tilapia, *Oreochromis niloticus* fed plant based low phosphorus diets supplemented with graded levels of different sources of microbial phytase. *Aquaculture*, 248: 111–119.

Liu, K., 1997. Chemistry and nutritional value of soybean components. In: *Soybeans: Chemistry, Technology, and Utilization*. Chapman and Hall, New York, pp. 25–113.

Mowrey, D., 1986. *The Scientific Validation of Herbal Medicine*, 187: 74–75.

Peter, R.E. and Crim, L.W., 1979. Reproductive endocrinology of fishes: Gonadal cycle and gonadotropin in teleosts. *Ann. Rev. Physiol.*, 41: 323–335.

Ray, J.W., 1990. 12 points on mature green papaya (pawpaw). Personal Publi.

Richardson, N.L., Higgs, D.A., Beams, R.M. and McBride, J.R., 1985. Influence of dietary calcium, phosphorus, zinc and sodium phytate level on cataract incidence, growth and histopathology in juvenile chinook salmon.

Rodehutscord, M. and Pfeffer, E., 1995. Effects of supplemental microbial phytase on phosphorus digestibility and utilization in Rainbow Trout (*Oncorhynchus mykiss*). *Water Sci. Technol.*, 31: 143–147.

Satoh, S., Poe, W.E. and Wilson, R.P., 1989. Effect of supplemental phytate and/or tricalcium phosphate on weight gain, feed efficiency and zinc content in vertebrae of channel catfish. *Aquaculture*, 80: 155–161.

Spinelli, J., Houle, C.R. and Wekell, J.C., 1983. The effect of phytase on the growth of rainbow trout (*Salmo gairdneri*) fed purified diets containing varying quantities of calcium and magnesium. *Aquaculture*, 30: 71–83.

Watt, J.M. and Breyer-Brandwijk, M.G., 1962. *The Medicinal and Poisonous Plants of Southern and Eastern Africa: Being an Account of their Medicinal and Other Uses, Chemical Composition, Pharmaceutical Effects and Toxicology in Man and Animal*, 2nd edn. E and S Livingstone Ltd., Edinburgh.

Chapter 8

On the Aspect of Seed Production and Prospects for Brackishwater Culture of Pearl Spot, *Etroplus suratensis*

☆ *S.D. Naik, S.T. Sharangdher, H.B. Dhamagaye and R.K. Sadawarte*

ABSTRACT

Etroplus suratensis is an euryhaline fish found in the backwaters of southern Maharashtra, Karnatka and Kerala state. It is herbivorous and being, cichlid has a well developed parental care. Its colouration and pearl spot markings make it a good aquarium fish in marine as well as freshwater. Being tasty, it makes a good table fish in all over India and hence enhances the prospects for culture practices along the west coast of India. Pearl spot culture can becomes as an alternative to shrimp culture during lean periods along the west coast of India. It is therefore to enhance its prospects in brackishwater aquaculture, its breeding and seed production technique was standardized in the laboratory condition. Observations made on its fry and fingerling rearing and their maturation etc are incorporated in this paper.

Introduction

Etroplus suratensis, the pearl spot is a naturally occurring cichlid fish in the back water of Indian coast. It has a restricted distribution along the south-west and south-east coast of India and Sri Lanka. Pearl spot, the only Asian brackishwater fish of India which can be easily made to spawn in confinement (Bardach, *et al.*, 1972 and Pannikkar, 1924). It is highly adopted to brackishwater systems. It is hardy fish but, breeds in impoundment shows parental care. This fish can be easily acclimatized to freshwater. Looking of its culture potential its commercial cultivation is still restricted only to the level of traditional farming/fishing in certain parts of country. This might be due to several reasons such as scanty/no supply of commercial seed, low fecundity, non availability of breeding stock. In nature, Pearl spot attach their eggs to the underside of the submerged objects at a depth of 1 meter or less in fresh or brackishwater. Culturist in southern India (the only area where Pearlspot are cultured) take advantage of this trait by erecting platform made of slabs of stone or slate in their ponds. Like all cichlid, Pearlspot take care of their young ones thus special spawning pond are not usually provided. Due to parental care exhibited in this species, larvae are not found in nature. Colouration of fish larvaes are found different from their adults. Pearlspot are mainly herbivore and feed mostly on blue and green algae and decaying plant remains. Pearlspot can attain a length of 10-12 cm in pond in one year and have been grow upto 20 cm.

Material and Methods

The collection of fry and adults were made at Ratnagiri by using dragnet mainly at the Government Brackishwater farm and Dr. BSKKV brackishwater farm, Zadgoan, Ratnagiri. The collections were made during the month of January and February. After collection, fry and adults of Pearlspot were kept in the quarantine tank for acclimatization. Treatment for injured specimen were given by bath of salt water first and then by 5 per cent formalin which was mainly for disinfection. After keeping in 3-4 days in quarantine tank, the fry and adult pearlspot were segregated according to their size. Measured the length and weight of all pearlspot fishes. Following three experiments were conducted in the laboratory.

Table 8.1: Fry Rearing Experiment

Sl.No.	*Parameters*	*Control*	*25% Protein Base Feed*	*40% Protein Base Feed*
1.	Volume of water in plastic pool	300 lit	300 lit	300 lit
2.	No.of fry stocked	30	30	30
3.	Initial avg. length (mm)	14.5	14.5	14.5
4.	Initial avg. weight (g)	0.056	0.056	0.056
5.	Salinity (ppt)	15.0	15.0	15.0
6.	Period (Days)	90	90	90
7.	Feed (Protein per cent)	Mixed zooplankton (10 per cent)	Artificial feed (25 per cent)	Artificial feed (40 per cent)
8.	Feeding rate	5 per cent	5 per cent	5 per cent
9.	Final avg. length (mm) after 90 days	32.63±1.24	40.25***±0.63	38.04***±0.53
10.	Final avg. weight(g) after 90 days	0.816±0.08	1.465***±0.07	1.290***±0.05
11.	Increase in length (mm)	18.13	25.75	23.54
12.	Increase in weight (g)	0.760	1.409	1.234
13.	Survival (per cent)	60	80	70
14.	FCR	0.55	0.22	0.29

***: $P < 0.05$±standard error of mean; ***: $P < 0.05$±standard error of mean.

Fry Rearing Experiment

For study the effect of different protein levels feed on the growth of pearl spot fry. Experiment was conducted on fry of Pearlspot (avg. 14.5 mm in length) by keeping 30 nos of fry @ 1 no/10 lit in the 3′ × 2′ plastic pool. Salinity of water kept at 15 ppt and mixed zooplankton feed was given as control. Pelleted feed were prepared by keeping the protein level 25 per cent and 40 per cent by using the locally available ingredients such as fish meal, algae, rice bran and groundnut oilcake. Feeding rate was kept as 5 per cent of the body weight during the experiment of 90 days. Monthly observation of length gain and weight gain of fry are recorded (Table 8.1).

Experiment on Effect of Hormones on the Growth of Fry

To stimulate the normal growth in Pearlspot fry, an experiment was conducted by incorporating growth promoting hormones like Methyl testosterone and Thyroxine in the Pelleted feed @ 1mg/kg, 3mg/kg, and 5mg/kg of fish. The feed pellets consisting of fish meal, algae, rice bran and groundnut oil cake. Duration of experiment was kept for 60 days. (Table 8.2).

Table 8.2: Effect of Methyl Testosterone and Thyroxine on Fry of Pearl Spot

	Control	*Methyl Testosteron*			*Thyroxine*		
Dose (mg/kg)	NIL	1 mg.	3 mg.	5 mg.	1 mg.	3 mg.	5 mg.
Days	60	60	60	60	60	60	60
No. of specimen used	30	30	30	30	30	30	30
Initial avg. wt. (gm.)	4.85	4.85	4.85	4.85	4.85	4.85	4.85
Final avg. wt. (mg.)	5.96	24.29	26.29	6.15	12.63	20.16	18.93
Avg. wt. gain (mg.)	10.81	29.14	31.14	11.00	7.78	15.31	14.08
Survival per cent	60	50	60	60	50	80	60

Maturation and Breeding Experiment

To study the maturation in adult, average size of avg. 132 mm and 49.16 gm of pearlspot specimen were reared in cement cistern for 90 days. Salinity of water was kept 15 ppt which was Brackishwater. To obtain the maturation among the adult, human chrionic gonadotrophin (HCG) was injected intramuscularly @ 10mg/kg of fish. Injection has to be given after every 30 days interval and fed with artificial feed (25 per cent protein). Another set was kept as a control where the adults were without injected but the feed was given on same feeding rate and time (Table 8.3). Experimental and control consist of six adult fish in each group. The plastic pool bottom was provided with farm soil and marine silt.

Observations

Fry Rearing Experiment

Average growth of fry after 90 days was found to be significantly higher among 25 per cent protein base feed and 40 per cent base feed as compared to that of control. But maximum gain in length and weight was observed in case of 25 per cent protein base feed (*i.e.* 25.75 mm and 1.409 gm in 90 days) than the 40 per cent of protein base feed.

Table 8.3: Breeding and Maturation Experiment

Sl.No.	Parameters	Control	Experimental
1.	Average length (mm)	132	132
2.	Salinity (ppt)	15.0	15.0
3.	Period (Days)	90	90
4.	Treatment (HCG mg/kg.)	Nil	10
5.	Initial weight (gm)	460	460
6.	Avg. weight gain in 30 days (gm)	7.0	9.34
7.	Avg. weight gain in 60 days (gm)	4.5	6.17
8.	Avg. weight gain in 90 days (gm)	3.17	3.66
9.	Gonado-somatic index (GSI)–Male	3.65 (± 0.06)	4.45 (± 0.63)
10.	Gonado-somatic index (GSI)–Male	0.127(± 0.06)	0.32 (± 0.007)

Experiment on Effect of Hormone on the Growth of Fry

After 90 days, there was an improvement in the weight gain among the methyl testosterone and thyroxine hormone treated as compared to untreated control group. Among these treatment, 3mg/kg thyroxine showed maximum weight gain (15.31 gm). Treatment of 5 mg/kg thyroxine was also give encouraging weight gain (14.08 gm) but less survival (60 per cent) as compared to 3 mg/kg treatment (80 per cent).

Breeding and Maturation

Experimental treatment (*i.e.* treated with HCG) male and female showed increased weight gain and higher GSI as compared to that of untreated controls. Gonado-somatic index was also observed in both sex. It was observed that in male 3.65 where control feed while maximum 4.45 in experimental feed. In case of female, GSI was observed 0.127 in control while maximum 0.320 in experimental. treatment. Among them selected three pairs of adul' Pearlspot of 46 gm in average weight were reared in 10′ × 2′ plastic pool which contained marine silt and farm soil as substratum. The fish attained the average weight gain 18.99 gm after 90 days in the HCG treated fish while untreated showed average weight gain only 14.22 gm. Natural breeding occurred during three month. From the first pair of female lay on an average 500 eggs on the adhesive substraturn. From that 200 fry were obtained under these condition. While from the second pair, 600 fry and 1300 nos. of fry were obtained from third pair of Pearlspot.

Results and Discussion

The Pearlspot is a naturally occurring cichlid fish in the Brackishwater of Indian coast. Experiments were conducted in this project to study the rate of growth, feed requirement, breeding and seed production of this fish so as to decide its suitability for brackishwater farming. Growth of Pearlspot fry to fingerling stage was studied under controlled conditions. Artificial feed with 25 per cent and 40 per cent level of protein was given to the fry to assess the protein requirement. It was observed that the survival of *E. suratensis* fry was significantly better in experimental feed as compared to the control. Among the experimental feed 25 per cent protein based feed showed maximum weight gain. There was no significant differences in growth with increasing dietary protein level from 25 per cent to 40 per cent but the effect was observed on survival percentage.

For the effect of hormones on the fry of pearlspot experiment showed that 3mg/kg thyroxine treatment gives the maximum growth and higher survival. Effect of Human Chorionic Gonadotrophin (HCG) was studied in adult Pearlspot for early maturation and breeding. The fishes were given HCG @ 10mg/kg injection intramuscularly. Fish were found mature at 90 days at the size of 150 mm size. Gonadosomatic index was analysed and found higher GSI in both sexes. The fish appeared to be responding favourably to HCG injection as seen from the higher GSI in male (4.45) and female (0.32) of treated fish as compared to untreated.

The pearlspot fish, thus was found to be tolerate brackishwater environment, accepts and grows on artificial diet, attains marketable size within a years culture period and can be induced for breeding in capacity. It is therefore, seems to be a suitable candidate species for introduction in brackish fish farming along the coast of Maharashtra.

Acknowledgement

We are indebted to Authority of Dr. B.S.Konkan Krishi Vidyapeeth for providing facilities to complete this work and we express our sincere gratitude to Dr. P.C.Raje, Associate Dean, College of Fisheries, Ratnagiri for encouragement in this work.

References

Bal, D.V and Rao, K.V., 1984. Mariculture. In: *Marine Fisheries.* Wiley-Interscience, John Wiley and Sons, London, 470 p.

Bardach, R.J., Ryther, J.H. and Mclaney, W.O., 1972. *Aquaculture.* Wiley Interscience, John Wiley and Sons, London, 79 p.

Dunnebier, R., 1932. Etroplus maculates pflege and zucht. *Aquar.–Terrienle,* 29(1): 4–5.

Jayram, K.C., 1981. In: *Handbook of Freshwater Fishes of India,* Director, Zoological Survey of Indian, Calcutta, 475 P.

Panikkar, N.P., 1921. Notes on the two cichlid fishes of Malbar, *Etroplus suratensis* and *Etroplus macalatus.* Bull, Madras Fish. Bur. No. 12 Rept. No. 5: 157–166.

Panikkar, N.P., 1924. Further notes on the breeding habits of the pearlspot fish (*Etroplus suratensis*). *J. Bombay. Nat. History Society,* 29(4): 1064.

Samarakorn, J.I., 1983. Breeding patterens of the indigenous cichlids *Etroplus suratensis* and *Etroplus macalatus* in an in Sri Lanka. *Mahesager Bull.,* 16(3): 357–362.

Raj, B.S., 1916. Note on the freshwater fish of Madras. *Rec. Indian Mus. Calcutta,* 12(17): 249–294.

Wickler, W., 1956. Der Haftapparat einiger cichliden Eier. *Zeitschr Zellforscti,* 45: 304–327.

Chapter 9

Food and Feeding Habits of Mudcrab, *Scylla* spp. of Ratnagiri Coast, Maharashtra

☆ *A.B. Funde, S.D. Naik, S.A. Mohite, G.N. Kulkarni and A.V. Deshmukh*

Introduction

The two crab species, *S. serrata* and *S. tranquebarica* are the most common occurring in the estuarine and mangrove areas along the west and east coasts of India. *S. serrata* (Forskal) is the only species of the family Portunidae that is closely associated with mangrove environments. Commonly known as the mangrove crab or mud crab, *S. serrata* is commonly called as "red crab" and it prefers to live in low saline waters, whereas *S. tranquebarica,* is the "green crab" living in high saline waters (Keenan *et al.,* 1998). Due to high demand and good price the species has been being overfished for the last one decade. At present mud crab is largely exported to Singapore, Hongkong and Malaysia. Due to its high demand in the world market in live condition, export of mud crab from India is increasing very rapidly recently. Both the crab species are generally estuarine, occurring both inter-tidally and sub-tidally. Although these are large prominent crustacean, little is known of its feeding behaviour and nothing has been published on its feeding habits (Hill 1976). It is commonly caught in traps baited with fish but it is not known whether fish forms part of its natural diet. In view of the crab farming concern in future, it would be very important to know the food of mudcrab at natural conditions. Therefore, a study on the food and feeding habbits of mudcrab *Scylla spp.* work was undertaken by collecting the local mud crab spp. from the Ratnagiri coast, Maharashtra.

Material and Methods

Samples were collected from different estuarine areas of Ratnagiri coast, Maharashtra namely Kasarweli, Shirgaon and Majgaon. Crab samples were caught by using the trap gear, locally known as '*Zella*' and other gill net. Species were identified and segregated on the basis of external morphological characters. Crab species samples were collected twice in a month during April 2007 to March 2008. Then collected crab spp. samples were dissected for observation of condition of gut. Remaining crab samples were kept for preservation in a deep freeze at –18°C prior for further analysis. Preserved sample were dissected within 24 hours.

A regular fortnightly collection was made either from gill net or trap landings (depending on availability) for investigating this aspect. Total number of 134 crabs of *S. serrata* and 97 crabs of *S. tranquebarica* having different sizes ranges (7 cm to 28.5 cm) carapace width were dissected for study of gut content and their analysis. These samples were collected during March-2007 to April-2008. After measuring carapace width, the crabs were either dissected washed thoroughly to make free of adhering foreign particles like sand, mud and other parts before the gut analysis. Gut analysis method was carried out as per the method described by the Hill (1976). The foregut content was identified to the lowest taxon under a compound microscope. The frequency of occurrence of materials determined by counting every foregut that contained at least one specimen of part of specific item taxon by (Prasad and Neelakantan, 1988). The content of gut volume was also observed before the crab sample dissected. The extent of food was determined by degree of distension of stomach and the amount of food contained in it. The condition of stomach was thus classified as full, 3/4 full, 1/2 full, 1/4 full and trace or empty as per reference work quoted by Pillay (1953), Sarojini (1954) and Luther (1962).

Observations

Observation were taken in three ways including fluctuations in feeding intensity, monthly composition of food items and annual composition of food items in *S. serrata* and *S. tranquebarica*. After the detailed observation throughout the year, follwing observation were recorded.

Fluctuation in Feeding Intensity

The gut content analysis of *S. serrata* and *S. tranquebarica* are presented in Table 9.1 and 9.2 respectively. Due to scarcity of specimens of *S. tranquebarica*, detailed gut analysis of this species could not be undertaken during the month of June to September. *S. serrata* exhibited only one intensive feeding peak in the year during the month of June to October (Table 9.1) and intensive peak feeding of *S. tranquebarica* was observed in October to December (Table 9.2).

Monthly Composition of Food Items in *S. serrata* and *S. tranquebarica*

S. serrata

Monthly percentages of five food components of *S. serrata* are presented in Table 9.3. It was found that crabs ingested major five food items namely crustacean matter, molluscan matter, fish matter, detritus matter and miscellaneous matter. Among these components crabs preferred detritus matter at maximum *i.e.* (71.47 per cent) in the month of July while fish matter maximum *i.e.* (35.27 per cent) in the month of May. Miscellaneous matter was more *i.e.* (34.21 per cent) in the month of March.

Table 9.1: Monthly Percentage of *Scylla serrata* in Various Condition of Feeding

Months	No of Guts Examined	Per cent of Stomach Condition				
		Full	3/4 Full	1/2 Full	1/4 Full	Trace
Apr-2007	13	5.69	9.70	46.15	23.08	15.38
May-2007	10	10.00	20.00	50.00	10.00	10.00
Jun-2007	11	**18.18**	**36.36**	27.27	9.09	9.09
Jul-2007	12	41.67	25.00	25.00	8.33	–
Aug-2007	10	20.00	40.00	20.00	20.00	–
Sept-2007	12	33.33	25.00	16.67	16.67	8.33
Oct-2007	11	**18.18**	**18.18**	36.36	–	27.27
Nov-2007	10	10.00	20.00	30.00	20.00	20.00
Dec-2007	13	15.38	23.08	–	38.46	23.08
Jan-2008	10	10.00	–	20.00	40.00	30.00
Feb-2008	10	–	–	20.00	50.00	30.00
Mar-2008	12	–	–	41.67	33.33	25.00

Table 9.2: Monthly Percentage of *Scylla tranquebarica* in Various Condition of Feeding

Months	No of Guts Examined	Per cent of Stomach Condition				
		Full	3/4 Full	1/2 Full	1/4 Full	Trace
Apr-2007	10	20	20	10	30	20
May-2007	11	18.18	36.36	27.27	9.09	9.09
Jun-2007	–	–	–	–	–	–
Jul-2007	–	–	–	–	–	–
Aug-2007	–	–	–	–	–	–
Sept-2007	–	–	–	–	–	–
Oct-2007	10	30	20	10	30	10
Nov-2007	10	-	50	20	20	10
Dec-2007	8	12.50	37.50	12.50	12.50	25.00
Jan-2008	5	-	20	20	40	20
Feb-2008	11	18.18	9.09	27.27	27.27	18.18
Mar-2008	10	30	30	10	20	10

S. tranquebarica

Monthly percentage of five food components occurred in *S. tranquebarica* are presented in Table 9.4. In case of the *S. tranquebarica*, it was observed that the five food components were same as found in *S. serrata* but occurrence of fish matter was found maximum *i.e.* (59.33 per cent) in the month of December, while detritus matter component maximum *i.e.* (35.38 per cent) in the month of October. The crustacean matter was found to be higher *i.e.* (28.57 per cent) in the month of April (Table 9.4).

Table 9.3: Monthly Percentage Composition of Food Items in *S. serrata* of Ratnagiri Coast

Months (year)	Food Items (per cent)					
	Crustaceans Matter	Molluscan Matter	Fish Matter	Detritus Matter	Miscellaneous Matter	Un-identified Item
Apr-2007	06.93	44.29	08.00	29.32	06.52	4.94
May-2007	22.07	10.00	35.27	14.74	16.00	1.92
Jun-2007	29.73	16.22	24.00	7.00	13.24	9.81
Jul-2007	3.57	7.33	02.35	71.47	13.43	1.85
Aug-2007	19.00	7.01	19.00	51.32	0.91	2.76
Sep-2007	16.30	13.00	9.27	39.83	12.52	9.08
Oct-2007	8.53	21.89	19.07	36.00	11.04	3.47
Nov-2007	23.64	17.00	04.27	19.86	32.54	2.69
Dec-2007	07.00	14.32	25.26	32.00	19.07	2.35
Jan-2008	13.71	3.07	34.00	28.30	13.68	7.24
Feb-2008	24.00	2.04	15.24	37.68	17.65	3.39
Mar-2008	9.14	13.24	17.00	22.00	34.21	4.41

Table 9.4: Monthly Percentage Composition of Food Items in *S. tranquebarica* of Ratnagiri Coast

Months (year)	Food Items (per cent)					
	Crustaceans Matter	Molluscan Matter	Fish Matter	Detritus Matter	Miscellaneous Matter	Un-identified Item
Apr-2007	28.57	8.35	35.66	21.3	3.9	2.22
May-2007	4.36	7.62	58.44	11.45	16.68	1.45
Jun-2007	–	–	–	–	–	–
Jul-2007	–	–	–	–	–	–
Aug-2007	–	–	–	–	–	–
Sep-2007	–	–	–	–	–	–
Oct-2007	1.71	14.53	39.49	35.38	6.50	2.39
Nov-2007	23.76	26	3.92	13.2	26.8	6.32
Dec-2007	15.83	6.17	59.33	2.67	11.83	4.17
Jan-2008	5.1	16.03	41.23	27.09	4.9	5.65
Feb-2008	25.93	6.76	25.99	12.66	25.29	3.37
Mar-2008	18.62	15.76	29.8	7.74	23.8	4.28

It was observed that the matter and detritus food component showed maximum contribution in the annual percentage of food item. Among these two, maximum fish matter was observed in case of *S. tranquebarica* while the maximum detritus matter was reported in *S. serrata*.

Table 9.5: Annual Percentage Composition of Food Items in *S. serrata* and *S. tranquebarica* of Ratnagiri Coast

Food Items	*Scylla serrata*	*Scylla tranquebarica*
Crustaceans matter	15.30	15.49
Molluscan matter	14.12	12.65
Fish matter	17.73	36.73
Detritus matter	32.46	16.44

Results and Discussion

Hill (1976, 1979), Prasad and Neelakantan (1988), Kneib and Weeks (1990) Chatterji *et al.* (1992), Nandi and Pramanik (1994) and Kathirvel *et al.* (1997) had described the food and feeding habit of *S. serrata* and *S. tranquebarica*. Feeding intensity of Indian horseshoe crab, *Tachypleus gigas* during different months was recorded by Chatterji *et al.* (1992). The information on feeding intensity of mud crab for different months were not recorded by any research worker, but the nature of food preferred was recorded. In the present work, higher feeding intensity were observed during June to September in case of *S. serrata* while it higher in the month of October to December. In case of *S. tranquebarica* this period may be the immature condition of the mud crab *Scylla* spp.

Hill, (1976) worked on the composition of food in the gut of *S. serrata*. He described the occurrence of molluscs (50 per cent), bivalve (30 per cent), and Crustacea (20 per cent) and very less volume of fishes in their gut. Prasad *et al.* (1988) noted detritus (35.70 per cent), fish (23.57 per cent) and crustacean (18.37 per cent) in the gut of *S. serrata*. Kneib and Weeks, (1990) working on mud crab *Eurytium limosum*, described the occurrence of some plant material, maximum amount of polycheats, bivalve and snails in the gut. Nandi and Pramanik, (1994) reported that the food of *S. serrata* consisted of detritus, crustacean and polychates. Kathirvel *et al.* (1997) reported that the food of *S. serrata* constituted by bivalve, mollusca, small crab and dead and decaying material.

In the present work the detritus matter (32.46 per cent) followed by fish matter (17.73 per cent), Miscellaneous matter (15.90 per cent) remains and crustacean matter (15.30 per cent) were observed in *S. serrata* while fish matter was maximum (36.73 per cent) followed by detritus matter (16.44 per cent) crustacean matter (15.49 per cent) in the *S. tranquebarica*. This observation supported the burrowing habit of the mud crab *S. serrata* found in estuarine areas while *S. tranquebarica* found mostly in deeper water that prefers fish food in their diet.

Acknowledgement

Authors are grateful to university authorities of Dr. B.S. Konkan Krishi Vidyapeeth, Dapoli for giving permission and providing necessary facilities to carryout these research work under research programme. We express our sincere gratitude to Dr. S.R. Kovale, Associate Dean, College of Fisheries, Ratnagiri for encouragement in this work.

References

Chatterji, A., Mishra, J.K. and Parulekar, A.H., 1992. Feeding behaviour and food selection in the horseshoe crab, Tachypleus gigas (Müller). *Hydrobiologia*, 246(1): 41–48.

Hill, B.J., 1976. Natural food foregut clearance-rate and activity of the crab *Scylla serrata*. *Marine Biology*, 34: 109–116.

Hill, B.J., 1979. Aspects of feeding strategy of the predatory crab *Scylla serrata*. *Marine Biology*, 55: 209–214.

Keenan, C.P., Davie, P.J.F. and Mann, D., 1998. A revision of the genus *Scyllu* de Haan, (1833 Crustacea: Decapoda: Brachyura: Portunidae). *Raffles Bulletin of Zoology*, 46: 217–245.

Kneib, R.T. and Weeks, C.A., 1990. Intertidal distribution and feeding habits of the mud crab, *Eurytium limosum*. *Estuarine Research Federation*, 13(4): 462–468.

Kathirvel, M., Srinivasagam, S., Ghosh, P.K. and Balasubramanian, C.P., 1997. *Mud Crab Culture*. CIBA Bulletin, Chennai, India, 10: 25.

Luther, G., 1962. The food habits of *Liza macrolepis* (Smith) and *Mugil cephalus* Lin. (Mugilidae). *Indian J. Fish.*, 9: 604–626.

Nandi, N.C. and Pramanik, S.K., 1994. *Crab and Crab Fisheries of Sundarban*. Hindustan Publishing Corporation, Delhi, India, pp. 192.

Pillay, T.V.R., 1953. *Mugil poecilus* Day same as *Mugil troscheli* Bleeker. *J. Bom. Nat. Hist. Soc.*, 51(2): 378–383.

Prasad, P.N. and Neelakantan, B., 1988. Morphometry of the mud crab *Scylla serrata*. *Seafood Export Journal*, 20(7): 19–22.

Sarojini, K.K., 1954. The food and feeding habits of the grey mullets *Mugil parsia* (Ham.) and *Mugil speigleri* (Bleeker). *Indian J. Fish.*, 1: 67–93.

Chapter 10

Potential for Ammonia and Nitrite Reducing Products for Shrimp Farms in Andhra Pradesh

☆ *N.A. Sadafule, S.S. Salim and A.D. Nakhawa*

ABSTRACT

The study indicated that the awareness of ammonia is very high among the group of farmers. They are also of opinion in its detrimental effect on the yield losses. Currently there exist as many 21 products in the market. The market continues to remain as monopolistic competition with more and more entrants in the market. There exist numerous products, sale services technical guidance, product promotion programs, incentives and credit availability to insure the brand loyalty of customer. Nevertheless there exists a scope for new product entry into the market. With more than 45 per cent of the customers showing their desire for the replacement of the product and 73 per cent of aspiring for the product with different futures and other services like consultancy farm delivery. Ammonia and nitrate problem in shrimp farming will continue to occur and based on the effects, perception, awareness visible symptoms as expressed by the different farmers should be taken into account.

Introduction

Fisheries and Aquaculture have been recognized as a powerful income and employment generator as it stimulates the growth of a number of subsidiary industries. It is an instrument for the livelihood for a large section of economically backward population of the country. Aquaculture is one of the fastest growing food sectors in the world with a impressive growth rate of over 8 per cent annually. Over the past three decades, there has been a rapid progress in aquaculture development all over the world, particularly in Asian countries. The major changes that are noticeable indicate the transformation from a small scale homestead level activity to a large scale commercial aqua farming. The inland fisheries sector in India contributes to about 50 per cent of the total fish production. During 2005-06 the contribution was 3.64 million tones. The shrimp contributes as one of the major commercially important species in aquaculture. Shrimp aquaculture amidst numerous governmental regulations, diseases occurrence, high cost of feed, the shrimp farm industry continuous to grow unabated. The farmed shrimp production in India during 2005-06 was 1,43,170 metric tones, which contributed to more than 80 per cent of the total shrimp production in the country.

Present Status of Fisheries and Aquaculture in Andhra Pradesh

Andhra Pradesh ranks first not only in shrimp and freshwater prawn production but also in costal aquaculture. Andhra Pradesh is one of the major states which contribute immensely to the fish production in the country; both inland as well as marine. It ranks second in inland fish production and fifth in marine fish production. Andhra Pradesh ranks second in production of value added fish products. Tiger shrimp, *Peneaus monodon* is the most important marine candidate species in India, which is farmed along the Indian coasts. Presently it is also farmed in freshwater systems mainly in East and West Godavari districts of Andhra Pradesh. The white spot syndrome virus disease out break, occurred along the East coast of India during 1994, spread through vertical and horizontal transmissions all along the Indian coasts. There are several shrimp farmers along the costal belt and the continuous release of WSSV affected waters from these farms increases the chances of spreading the disease and cause huge economic loss to the farmers.Though scampi farming has been seen as an alternative to tiger shrimp farming, the same could not share the popularity due to its inherent disadvantage which includes poor growth rate, labour intensiveness, and lack of quality seed material in time as well as in quantity, etc. Godavari is a perennial river and supplies water throughout the year from its canal system. The groundwater of Godavari districts is saline in nature (20 to 30 ppt). Therefore the farmers fulfill their need of saltwater during shrimp farming by pumping the groundwater. Shrimp farming in this region is being carried out at a salinity range between 1 to 15 ppt. The size of ponds used for shrimp farming in this region varies widely between 1 to 10 acres, while majority of the farmers have an area of 4 acres. The soil is clayey and alkaline in nature and has good water holding capacity which enables the farmers to adopt zero water exchange, though minimal water exchange is not uncommon.

Need for the Study

The ammonia and nitrate problems are considered to be detrimental to the sustainability of aquaculture industry in Andhra Pradesh, it is important to know about the awareness and technical knowledge about the possible symptoms, mitigating measures and quantification of losses. In addition, the farmers are very much pricing responsive. The Ammonia and Nitrite reducing products is a growing market on account of the capital intensive nature of shrimp farming. There exist many products in the market. It is also important to know about the different competing product, prices, technical support provided. The market potential for the different ammonia and nitrite reducing products was done in the state of Andhra Pradesh with the following objectives.

Objectives

- ☆ To estimate the economic loss due to the Ammonia and Nitrite.
- ☆ To analyse the awareness about different products available in Andhra Pradesh market.
- ☆ To develop a marketing strategy for the new entering product in the market.

Materials and Methods

The aim of the study was to find out the awareness and economic loss due to the ammonia and nitrite and to develop a viable marketing strategy for a new product entry in the market in selected districts of Andhra Pradesh. A structured questionnaire was developed which contained details on the personal information, awareness and causes of Ammonia and Nitrite, suggestion for product development, presently using products, any indigenous methods, about test kits used by them, about danger, action and acceptable levels of above said problems and awareness about use of chemicals and eco-friendly products.

Study Area

Andhra Pradesh was selected purposively for the study as Andhra Pradesh ranks first in shrimp and prawn production. The culture practices in Andhra Pradesh are semi intensive and intensive with high stocking density and excessive feeding. Due to high stocking density problems like Ammonia, Nitrite, are unavoidable in the farm. Lot of shrimp area was left abandoned or shrimp farming closed down as a result of the yield losses. The sampling was done in selected district of Andhra Pradesh *viz.*, East Godavari, West Godavari and Nellore. From each districts, 30 respondents were selected as big, small, and medium while care was taken that the selected respondents were progressive.

Analytical Tools

The data was collected using a structured questionnaire including the different aspects of shrimp farming, and with specific reference to ammonia and nitrate problem. The data was analyzed using average and percentage analyses which were supplemented by suing graphical illustrations have been used for analyzing the data.

Results and Discussion

The detailed survey schedule was employed to collect the different qualitative and quantitative information related to shrimp farming. The details on different problems related to Ammonia and Nitrite, were collected on account of its awareness, causes, effects, reducing products, etc. Numerous average and percentage analysis were estimated and graphical illustration were done. In order to deduce meaningful conclusions, the data were stratified into four namely, marginal, small, medium, and big farmers. In the context of the study, the marginal farmer has a land holding of less than 5 acre farm, small farmer 5-10 acre, medium farmer 10-20 acre and big farmer more than 20 acre farm area. For better clarity and understanding the results and discussion on the different Ammonia and Nitrite reducing products are discussed under the following heads. The details about the perception and awareness about ammonia and nitrite and their possible causes are furnished in Table 10.1. The major causes suggested were excess feeding, rise in temperature, rise in pH, and high stocking density.

The analysis reveals that 84.4 per cent of the total respondents are aware about ammonia problem and 33.3 per cent about nitrite problem. In East Godavari district, awareness about ammonia is 93.3 per cent and only 13.3 per cent were aware about nitrite. Most of the farmers feel that excess feeding (93 per cent) and high stoking density (86.6 per cent) were the major causes of ammonia and nitrite problem. In West Godavari district, awareness about ammonia is 86.6 per cent and nitrite was 53.3 per cent. Majority of the farmers consider that excess feeding (90 per cent) and high stoking density (73.3 per cent) and rise in pH (73.3 per cent) are the major causes of ammonia and nitrite problem. In Nellore district, awareness about ammonia was 73.3 per cent and nitrite was 33.3 per cent. Most of the farmers feel that excess feeding (83.3 per cent) is responsible for ammonia and nitrite problem. The analysis also reveals that the awareness level about ammonia and nitrite problems is highest among big farmers (100 per cent) and lowest among the marginal farmers (62.28 per cent). This indicates its higher occurrence and visible symptoms in big farms. The details about the Perception and Awareness about the effects of Ammonia and Nitrite problem are furnished in Table 10.2. The major effects are feed drop, loose shell problem, black gill formation, reddish pleopods and periopods, reddish body colour reddish, etc.

Table 10.1: Perception and Awareness about Ammonia and Nitrite problem and its possible causes

Sl.No.	*District*	*Farm Size*	*AAA*	*AAN*	*EFD*	*RTM*	*RpH*	*HSD*
1.	EG	0-5	8 (80)	0 (0)	8 (80)	6 (60)	7 (70)	7 (70)
		5-10	8 (80)	0 (0)	8 (80)	6 (60)	7 (70)	7 (70)
		10-20	6 (100)	0 (0)	6 (100)	6 (100)	5 (83.3)	6 (100)
		> 20	6 (100)	2(33.3)	6 (100)	6 (100)	6 (100)	6 (100)
		Total	28(93.3)	4(13.3)	28(93.3)	26(86.6)	24(80)	26(86.6)
2.	WG	0-5	5 (62.5)	1 (12.5)	7 (87.5)	5 (62.5)	5 (62.5)	3 (37.5)
		5-10	6 (85.7)	4(57.1)	6(85.7)	5 (71.4)	4(57.1)	5(71.42)
		10-20	7 (87.5)	4(50)	6(75)	7 (87.5)	6 (75)	7 (87.5)
		> 20	8 (100)	7 (87.5)	8 (100)	8 (100)	7 (87.5)	7 (87.5)
		Total	26(86.6)	16(53.3)	27(90)	25(83.3)	22(73.3)	22(73.3)
3.	NEL	0-5	5 (50)	0 (0)	7 (70)	1 (10)	2 (20)	3 (30)
		5-10	11 (84.6)	5(38.4)	11(84.6)	5(38.4)	8(61.5)	9 (69.23)
		10-20	6 (85.7)	5(71.4)	7(100)	7(100)	6(85.7)	5(71.42)
		Total	22(73.3)	10(33.3)	25(83.3)	13(43.3)	16(56.3)	17(56.6)
4.	TOT	0-5	18(62.2)	1(3.57)	22(78.57)	12(42.85)	14(50)	13(46.42)
		5-10	25(89.2)	11(39.2)	25(89.28)	18(64.28)	18(64.28)	21 (75)
		10-20	17(94.4)	8 (44.4)	17(94.4)	18 (100)	15 (83.3)	16(88.88)
		> 20	16 (100)	10 (62.5)	16 (100)	16 (100)	15 (93.7)	15(93.75)
		Total	76(84.4)	30(33.3)	80(88.8)	64 (71.1)	62 (68.8)	65(72.2)

*: Figures in parentheses indicates percentage to total.

AAA: Awareness about Ammonia; AAN: Awareness about Nitrite; EFD: Excess feeding; RTM: Rise in temperature; RpH: Rise in pH; HSD: High stocking density.

The analysis reveals that loose shell problem (74.4 per cent) and reddish body colour (78.9 per cent) are the most visible effects of ammonia and nitrite problem. In East Godavari, 70 per cent farmers are able to recognize that reddish body colour of the shrimps is because of stress on the shrimps due to ammonia and 60 per cent farmers were aware about the problems of black gill formation due to accumulation of ammonia and nitrite in the gills. In West Godavari, majority of the farmers were facing problems like reddish pleopods (96.6 per cent) and reddish body colour (96.6 per cent) of the shrimp. Farmers in Nellore district consider that effects of ammonia and nitrite are reddish pleopods (80 per cent), feed drop (76 per cent) and loose shell problem (76 per cent).The analysis reveled that farmers are aware that ammonia and nitrite in pond is responsible for reddish body colour of shrimp. The details about the perception and Awareness of the respondents on the yield loss due to ammonia and nitrite are furnished in Table 10.3. Due to excess ammonia and nitrite, shrimp gets stressed and it shows lethargic behavior and at last mortality occurs.

Table 10.2: Perception and Awareness about the effects of Ammonia and Nitrite problem

Sl.No.	District	Farm Size	PPR	BCR	FD	LSP	Ph	BGF
1.	EG	0-5	4 (40)	7 (70)	3 (30)	5 (50)	8.4	6 (60)
		5-10	3 (37.5)	5 (62.5)	5 (62.5)	4 (50)	8.4	4 (50)
		10-20	3 (50)	3 (50)	2 (33.3)	3(50)	8.3	2 (33.3)
		> 20	0 (0)	6 (100)	1 (16.6)	5 (83.3)	8.6	6 (100)
		Total	10(33.3)	21(70)	11(36.6)	17(56.6)	8.4	18(60)
2.	WG	0-5	8 (100)	8 (100)	6 (75)	7 (87.5)	8.3	3 (37.5)
		5-10	7 (100)	3 (100)	7 (100)	7 (100)	8.3	2 (28.5)
		10-20	6 (75)	6 (75)	5 (62.5)	5 (62.5)	7.2	3 (37.5)
		> 20	8 (100)	8 (100)	6 (75)	8 (100)	8.3	7 (87.5)
		Total	29(96.6)	29(96.6)	24(80)	27(90)	8.0	15(50)
3.	NEL	0-5	7 (70)	5 (50)	7 (70)	9 (90)	8.5	6 (60)
		5-10	1(76.2)	10 (76.9)	10(76.9)	7(53.8)	6.5	5 (38.4)
		10-20	7 (100)	6 (85.71)	6 (85.7)	7 (100)	6.9	6 (85.71)
		Total	24(80)	21(70)	23(76.6)	23(76.6)	7.3	17(56.6)
4.	TOT	0-5	19(67.8)	20 (71.4)	16(57.1)	21 (75)	8.4	15(53.5)
		5-10	20(71.4)	22 (78.5)	22(78.5)	18(64.2)	7.5	11(39.2)
		10-20	15(83.3)	13 (72.2)	12(66.6)	13(72.2)	7.7	9(50)
		> 20	9 (56.2)	16(100)	8 (50)	15(93.7)	8.4	15(93.7)
		Total	63(70)	71(78.9)	58(64.4)	67(74.4)	7.9	50(55.6)

*: Figures in parentheses indicates percentage to total.

PPR: Pleopods and periopods reddish; BCR: Body colour reddish; FD–Feed drop; LSP: Loose shell problem; pH: pH of Water; BGF: Black gill formation.

The analysis reveals that in general 87.8 per cent of the farmers are aware that ammonia problems lead to loss of yield and this problem is encountered on the 54th day of the culture period on an average. Individually this awareness in East Godavari, Nellore and West Godavari is 76.6 per cent, 93.3 per cent and 93.3 per cent respectively. The problem was generally encountered on 42nd, 53rd and 45th day of the culture period in East Godavari, Nellore and West Godavari districts respectively. Analysis reveals that most of the farmers are accept that there is loss due to ammonia and nitrate. There are number of ammonia and nitrite reducing products used in the Andhra aquaculture market. Deodarase, is having major market share (about 20 per cent) among all these products followed by Odoban, Proxy PS, and Spark PS etc. The results revealed that 24.4 per cent of the farmers are getting farm delivery and 6.7 per cent of the farmers are provided consultancy with the product they purchase. From the study, it has been seen that in East Godavari, 30 per cent farmers are getting farm delivery and 26 per cent farmers in West Godavari. In West Godavari only 10 per cent farmers are getting the facility of consultancy for the ammonia and nitrite reducing product. In Nellore 26.6 per cent farmers are getting farm delivery and 10 per cent are getting consultancy. It has been seen that generally the big farmers are provided this type of facilities because they gave the orders in bulk on account of. Small farmers are not able to get farm delivery and consultancy services. The respondent opinion on the product services namely farm delivery and consultancy are listed in Table 10.4. During the survey questions are asked to the farmers about satisfaction with product, replacement of the product, features desirable in the new product like low price, good results, farm delivery etc. The study reveals that out of 90 respondents, 73.3 per cent are satisfied with the current product, 45.6 per cent farmers want to replace the current product, and 73.3 per cent farmers want a new product with different features as stated above. In East Godavari, 63.3 per cent farmers are satisfied with their currently using product, 50 per cent farmers want to replace the currently using product and 73.3 per cent farmers want a new product. In West Godavari district, 70 per cent farmers are satisfied, 43.3 per cent farmers want to replace their currently using product and 73.3 per cent farmers want a new product. In Nellore district, 86.6 per cent farmers are satisfied, 45.6 per cent farmers want to replace their currently using product and 73.3 per cent farmers want a new product. It has been seen that generally the small farmers want to replace the current product so they can be targeted by a new entrant.

Table 10.3: Perception and Awareness of the Respondents on the Yield Loss due to Ammonia

Sl.No.	*District*	*Farm Size*	*LOS*	*TIO*
1.	EG	0-5	6 (60)	46
		5-10	6 (75)	61
		10-20	5 (83.3)	54
		> 20	6 (100)	70
		Total	23(76.6)	58
2.	WG	0-5	6 (75)	42
		5-10	7 (100)	61
		10-20	7 (87.5)	46
		> 20	8 (100)	62
		Total	28(93.3)	53
3.	NEL	0-5	10(100)	24
		5-10	11(84.6)	50
		10-20	7 (100)	60
		Total	28(93.3)	44
4	TOT	0-5	22(78.5)	37
		5-10	24(85.7)	56
		10-20	17(94.4)	53
		> 20	16(100)	66
		Total	79(87.8)	53

*: Figures in parentheses indicates percentage to total.

LOS: Awareness about loss due to Ammonia; TIO: Time of incidence occurrence (days onwards).

Table 10.4: Respondent Opinion About Farm Delivery and Consultancy

Sl.No.	*District*	*Farm Size*	*FAD*	*CON*
1.	EG	0-5	0 (0)	0 (0)
		5-10	2 (25)	0 (0)
		10-20	2 (33.3)	0 (0)
		> 20	5 (83.3)	0 (0)
		Total	9 (30)	0 (0)
2.	WG	0-5	2 (0)	0 (0)
		5-10	0 (0)	1 (14.2)
		10-20	0 (0)	0 (0)
		> 20	8 (100)	2 (25)
		Total	8 (26.6)	3 (10)
3.	NEL	0-5	0 (0)	0 (0)
		5-10	1 (7.6)	2 (15.3)
		10-20	4 (57.1)	1 (14.2)
		Total	5 (16.6)	3 (10)
4.	TOT	0-5	0 (0)	0 (0)
		5-10	3 (10.7)	3 (10.7)
		10-20	5 (27.7)	1 (5.5)
		> 20	14(87.5)	2 (12.5)
		Total	22(24.4)	6 (6.7)

*: Figures in parentheses indicates percentage to total.

FAD: Farm delivery; CON: Consultancy; ASS: After sales services.

Conclusions

The major findings which had emanated from the study are listed below:

1. Awareness of ammonia is 85 per cent and nitrite is 33 per cent in the sampled area.
2. Most of the farmers consider that both ammonia and nitrite problems are one and the same.
3. Most visible effects of ammonia and nitrite are reddish body colour and loose shells.
4. The Incidence of occurrence of the problem is on the 58th day in EG, 53rd day in WG and 44th day in Nellore district.
5. Only big farmers are getting the services like farm delivery and consultancy.
6. 73 per cent farmers are satisfied with currently using product because they are thinking that they are using best product in the market.
7. 46 per cent farmers want to replace the current product.
8. 73 per cent farmers wanted product having lesser price, good and fast results.

9. Mostly the big farmers are satisfied with the products they are using while a major proportion of marginal and small farmers want to replace the product.
10. Deodrase has maximum market share in ammonia and nitrate reducing products fallowed by Odoban, Proxy PS, and Spark PS.

References

Anon. Dramatic drop in farmed shrimp production, Andhra Pradesh. *Fishing Chimes*, 7(1): 127–135.

Anon. *Training Manual on Recent Advances in Management of Water Quality Parameters in Aquaculture.* Central Institute of Fisheries Education, Mumbai.

Athithan, S., 2007. Earthy odour and its effects in highly manure loaded and pellet fed fish pond. *Aqua International.*

Ayyappan, S., 2005. Sustainability of marine fisheries of India issues and approaches. *Fishing Chimes*, 25(1).

Camargo, Julio A., 2006. Ecological effects of inorganic Nitrogen pollution in aquatic ecosystems: A global assessment. *Environment International*, 32: 831–849.

Dwivedi, S.N., Saxena, Ankush and Bajpal, Avinash, 2005. Inland fisheries of India and future prospective. *Fishing Chimes*, 25(1).

Handbook of Fisheries and Aquaculture, 2006. Directorate of Information and Publications for Agriculture, Indian Council for Agricultural Research, New Delhi, 2006 Action plan MPEDA (2005–06).

Kumar Yashwant, J., 2007. Managing Ammonia in fish ponds. *Aqua International.*

Sekar, M., Krishnan, P. and Reddy, A.K., 2006. Tiger shrimp farming in freshwater system of Godavari districts, A.P., Management issues. *Fishing Chimes*, 26(1): 56–58.

Suresh, K., 2007. Probiotics: The boost for probiotics in aqua feeds. *Aqua International.*

Chapter 11

Evaluation of Effect of Different Salinities on Growth and Maturity of Indian Medaka, *Oryzias melastigma* (Mcclelland, 1839)

☆ *A.S. Pawar, S.T. Indulkar and B.R. Chavan*

Introduction

The Indian medaka is extremely adaptable fish and thrives under wide variety of conditions. It is an oviparous fish, reproduces through external fertilization. Because of this, direct observation of all the stages of early development of embryos, outside the mother is possible. Adult medaka grows maximum up to 4.0 cm in length (Talwar and Jhingran, 1991), can be easily maintained in aquarium tanks. It breed readily in captivity and have short generation time, small cultures can provide enough embryos for various types of research. These qualities of medaka make it a useful research animal (Kirchen and West, 1976). Recently, medaka is used as a sentinel or biomarker for environmental contamination by aquatic and environment toxicologist (Denny *et al.*, 1991). The embryo and just hatched larvae are the most delicate stages, very sensitive towards the environmental changes.

The Indian medaka being a euryhaline fish, can tolerate high degree of salinity from zero to 31 gL^{-1} (Manna, 1989). The salinity plays an important role in its maturity, reproduction, survival, metabolism and distribution. Changes in salinity or temperature may affect the maturation, spawning, fertilization etc. Knowledge of the pattern of early larval development and the effects of environmental parameters on reproduction is important as it facilitates aquaculture research and fish resource management. The possibility of their successful growth and reproduction in diverse aquatic salinity cannot be investigated in nature but only through the breeding experiment under laboratory conditions.

The research on Indian medaka, *O. melastigma* is at primitive stage. Hence, the present study was undertaken to see the effect of different salinities on growth and maturity of Indian medaka, *O. melastigma*.

Material and Methods

In the present study, 20 days old young ones (14.0±0.05, 10.5±0.03) produced from the same batch were used for the experiment. For first 20 days, these young ones were reared in freshwater and then gradually transferred to experimental conditions as freshwater (T_1), 5 (T_2), 10 (T_3), 15 (T_4), 20 (T_5), 25 (T_6), 30 (T_7) and 35 gL^{-1} (T_8) salinities. Experiment was conducted in plastic tubs (10 L capacity each) following completely randomised design with eight treatments having three replicates for each treatment. Fishes were stocked at six numbers per replicate and fed *ad libitum* twice per day with flake feed. Uneaten feed and faeces were siphoned out and nearly 30 percent of water from each tank was exchanged daily. Observations on weight gain, length gain, percentage weight gain, percentage length gain, specific growth rate, survival and Gonado somatic index (GSI) were made after 90 days of rearing period.

All data (length, weight, specific growth rate, survival and Gonado Somatic Index) were presented as mean±standard error of mean. The influence of salinity on all of the above indices was analysed by One-way Analysis Of Variance (ANOVA). Significant difference was tested at 0.05 level. To find out the significant difference among the treatments, Student-Newman-Keuls (SNK) multiple comparison test was used. Standard Error (SE) of the mean length, weight, and specific growth rate was calculated by following standard methods (Snedecor and Cohran, 1967; Zar, 2005).

Results

Observations on growth and survival of *O. melastigma* reared in different salinities from 0 to 35 gL^{-1} at an interval of 5 gL^{-1} salinity for 90 days are given in Table 11.1.

The average length gain (per cent) of fish was 112.38±0.31, 113.93±1.15, 153.69±0.86, 151.98±0.10, 144.76±0.83, 143.12±0.54, 132.62±0.93 and 122.50±0.55 per cent reared in treatments T_1, T_2, T_3, T_4, T_5, T_6, T_7, and T_8 respectively. ANOVA showed significant difference ($P<0.05$) in length gain of fish. The maximum length gain (153.69 per cent) was observed in the treatment T_3, whereas, minimum length gain (112.38 per cent) was observed in the treatment T_1 (Figure 11.1).

Student-Newman-Keuls multiple range test revealed that the maximum length gain was recorded in T_3, which differed significantly ($P<0.05$) with treatments T_1, T_2, T_5, T_6, T_7, and T_8 except with treatment T_4. Length gain (per cent) in T_5 did not differ significantly ($P>0.05$) with T_6, similarly T_2 did not differ ($P>0.05$) significantly with T_1. Treatments T_5 and T_2 differed significantly ($P<0.05$) with all other treatments.

The mean weight gain (per cent) was 2582.54±46.20, 2614.29±14.29, 3625.40±13.84, 3576.19±13.75, 3479.37±17.46, 3430.16±11.45, 3180.95±28.57 and 3084.13±15.14 per cent reared in treatments T_1, T_2, T_3, T_4, T_5, T_6, T_7, and T_8 respectively. ANOVA showed significant difference ($P<0.05$) in percentage gain in weight reared in different salinities. The percentage weight gain was higher (3625.40 per cent) in treatment T_3, whereas, it was less (2582.54 per cent) in treatment T_1 (Figure 11.2)

Student-Newman-Keuls multiple range test revealed that significantly maximum weight gain was observed in T_3 and it differed significantly ($P<0.05$) with other treatments except T_4. Weight gain (per cent) of fish in treatment T_5 differed significantly ($P<0.05$) with all other treatments except T_6. Similarly, the weight gain (per cent) in T_1 differed significantly ($P<0.05$) with all other treatments except T_2.

Figure 11.1: Length Gain (per cent) of *O. melastigma* Reared in Different Salinities for 90 Days

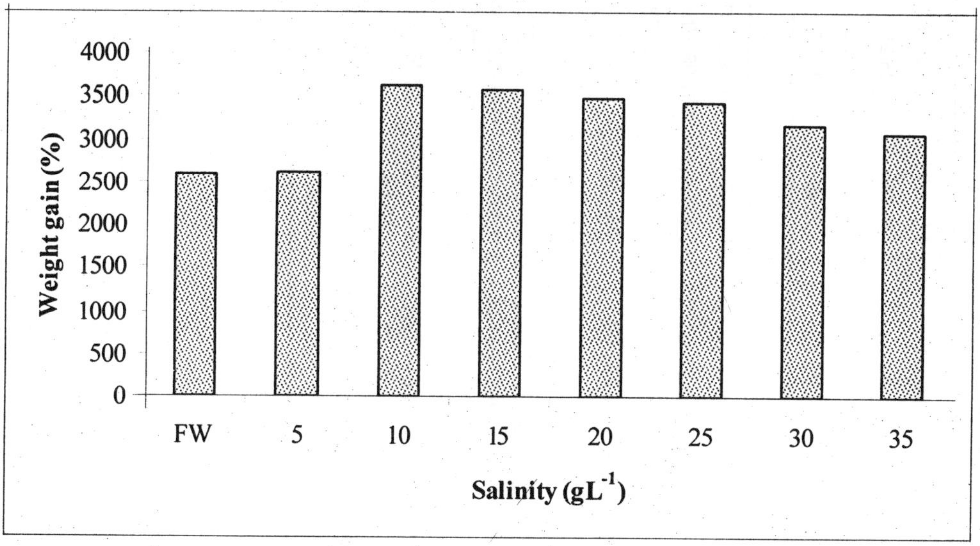

Figure 11.2: Weight Gain (per cent) of *O. melastigma* Reared in Different Salinities for 90 Days

The specific growth rate was 3.66±0.0197, 3.67±0.0052, 4.02±0.0041, 4.00±0.0042, 3.98±0.0054, 3.96±0.0036, 3.88±0.0097 and 3.85±0.0053 per cent day^{-1} reared in treatments T_1, T_2, T_3, T_4, T_5, T_6, T_7, and T_8 respectively. ANOVA showed significant difference ($P<0.05$) in specific growth rate reared in different salinities. The maximum specific growth rate (4.02 per cent day^{-1}) was observed in T_3 whereas, minimum SGR (3.66 per cent day^{-1}) was observed in T_1 (Figure 11.3).

Table 11.1: Data on Growth and Survival of *O. melastigma* Reared in Different Salinities for 90 Days

Particulars		Treatments (Salinities, gL^{-1})							
		T_1 (FW)	T_2 (05)	T_3 (10)	T_4 (15)	T_5 (20)	T_6 (25)	T_7 (30)	T_8 (35)
Initial length (mm)		14.0±0.05	14.0±0.05	14.0±0.05	14.0±0.05	14.0±0.05	14.0±0.05	14.0±0.05	14.0±0.05
Length after 90 days (mm)	R_1	29.80	30.00	35.75	35.28	34.50	33.90	32.80	31.30
	R_2	29.65	30.20	35.35	35.30	34.15	34.05	32.35	31.05
	R_3	29.75	29.65	35.45	35.25	34.15	33.90	32.55	31.10
Mean		29.73±0.04	29.95±0.16	35.52±0.12	35.28±0.01	34.27±0.12	33.95±0.05	32.57±0.13	31.15±0.08
Length gain (mm)		15.73±0.04	15.95±0.16	21.52±0.12	21.28±0.01	20.27±0.12	19.95±0.05	18.57±0.13	17.15±0.08
Length gain (%)		112.38^a±0.31	113.93^a±1.15	153.69^e±0.86	151.98^e±0.10	144.76^d±0.83	143.12^d±0.54	132.62^c±0.93	122.50^b±0.55
Initial weight (mg)		10.5±0.03	10.5±0.03	10.5±0.03	10.5±0.03	10.5±0.03	10.5±0.03	10.5±0.03	10.5±0.03
Weight after 90 days (mg)	R_1	272.50	282.00	393.50	386.00	379.50	369.00	347.50	331.50
	R_2	283.50	286.50	388.50	388.50	374.00	370.00	347.50	334.50
	R_3	289.00	286.50	391.50	383.50	374.00	373.00	338.50	337.00
Mean		281.67±4.85	285.00±1.50	391.17±1.45	386.00±1.44	375.83±1.83	370.67±1.20	344.50±3.00	334.33±1.59
Weight gain (mg)		271.17±4.85	274.50±1.50	380.67±1.45	375.50±1.44	365.33±1.83	360.17±1.20	334.00±3.00	323.83±1.59
Weight gain (%)		2582.54^a± 46.20	2614.29^a± 14.29	3625.40^e± 13.84	3576.19^e± 13.75	3479.37^d± 17.46	3430.16^d± 11.45	3180.95^c± 28.57	3084.13^b± 15.14
SGR (% day^{-1})		3.66^a±0.0197	3.67^a±0.0052	4.02^e±0.0041	4.00^e±0.0042	3.98^d±0.0054	3.96^d±0.0036	3.88^c±0.0097	3.85^b±0.0053
Survival		100	100	100	100	100	100	100	100

a, b, c, d, e: Values in a row sharing similar letter do not differ (P>0.05).

±: Standard Error of mean.

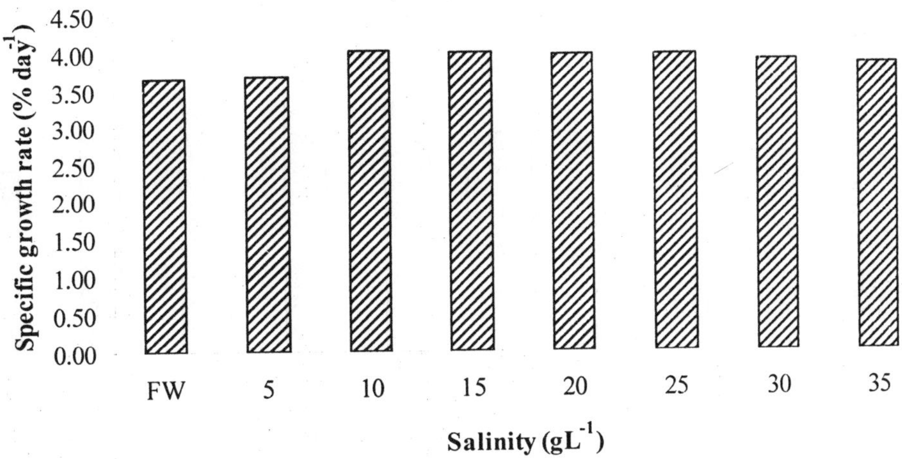

Figure 11.3: Specific Growth Rate (per cent day^{-1}) of *O. melastigma* Reared in Different Salinities for 90 Days

Student-Newman-Keuls multiple range test revealed that significantly maximum specific growth rate was observed in T_3, which differed significantly (P<0.05) with other treatments except with treatment T_4. Specific growth rate in T_5 did not differ significantly with T_6. Treatment T_1 did not differ significantly with T_2, and these treatments T_1 and T_5 differed significantly from all other treatments.

Cent-per-cent survival of fishes in all replicates of each treatment was observed after 90 days of rearing period.

The average GSI of male was 1.07±0.01, 1.30±0.01, 2.62±0.01, 2.61±0.01, 2.48±0.01, 2.32±0.01, 1.98±0.02 and 1.60±0.01 in treatments T_1, T_2, T_3, T_4, T_5, T_6, T_7 and T_8 respectively. ANOVA showed significant difference (P<0.05) in GSI of males reared in different salinities. The GSI of male was higher (2.26) in T_3 whereas, lower (1.07) in T_1 (Figure 11.4).

Student-Newman-Keuls multiple range test revealed that significantly maximum GSI (2.62) was observed in T_3, which differed significantly (P<0.05) from other treatments, except T_4. Significantly minimum GSI (1.07) was observed in T_1 which differed significantly from all other treatments.

The GSI of female in T_1, T_2, T_3, T_4, T_5, T_6, T_7 and T_8 was 1.83±0.03, 1.84±0.02, 3.24±0.02, 3.22±0.02, 2.73±0.01, 2.60±0.03, 2.40±0.02 and 2.37±0.01 respectively. ANOVA showed significant difference (P<0.05) in GSI of females reared in different salinities. It was higher (3.24) in T_3, whereas it was very lower in T_1 (1.83) (Figure 11.5).

Student-Newman-Keuls multiple range test revealed that significantly higher GSI was observed in T_3 and it differed significantly from treatments T_1, T_2, T_5, T_6, T_7 and T_8 except T_4. GSI in T_7 and T_8 did not differ significantly. Significantly lower GSI of female was observed in T_1 and it did not differ significantly from T_2.

During 90 days rearing period, temperature varied in the range of 26.8 to 31.5 °C with an average of 29.0 to 29.3 °C. The pH in the range of 7.25 to 8.20 with an average of 7.42 to 8.11, dissolved oxygen

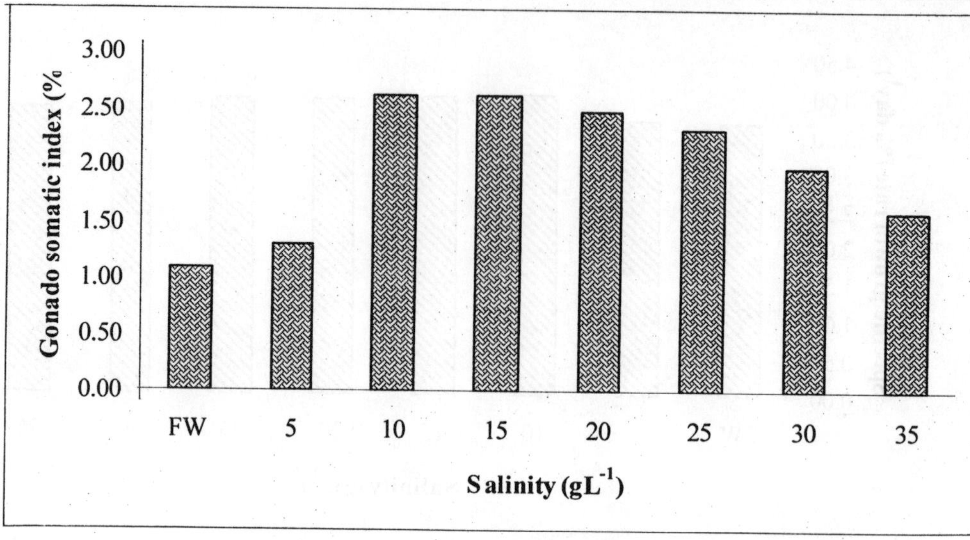

Figure 11.4: Gonado Somatic Index of *O. melastigma* (Male) Reared in Different Salinities for 90 Days

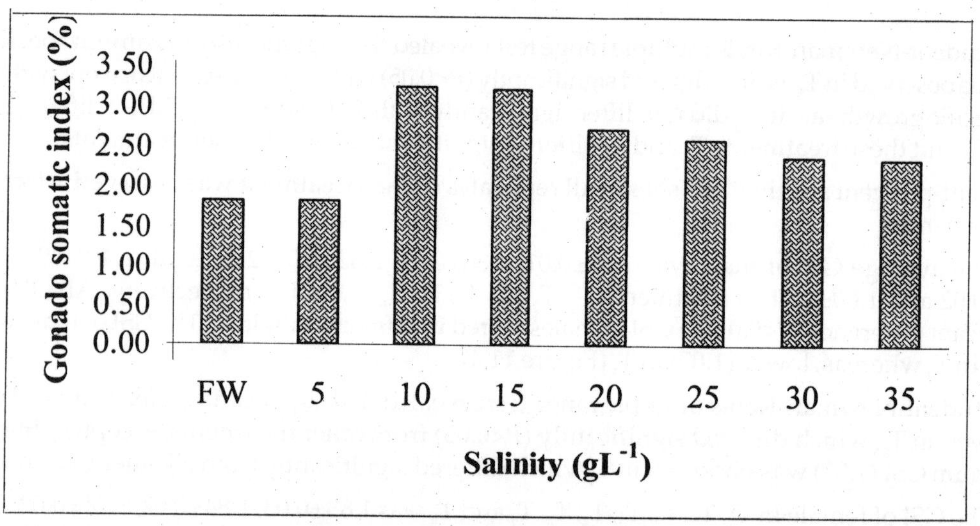

Figure 11.5: Gonado Somatic Index of *O. melastigma* (Female) Reared in Different Salinities for 90 Days

varied in the range of 5.2 to 8.0 mgL^{-1} with an average of 6.2 to 6.7 mgL^{-1} and total alkalinity in the range of 145 to 239 mgL^{-1} with an average of 190 to 216 mgL^{-1}. During 90 days of experimental period, the water parameters in all the treatments were almost same and were maintained within the optimum range.

From the forgoing account and considering the average length and weight gain, specific growth rate combine for male and female and separately for both and Gonado somatic index separately for male and female, 10 to 15 gL^{-1} salinity was found to be the most suitable and optimum salinity range for maximum growth and early maturation of *O. melastigma* fish.

Discussion

During the present study maximum length gain for male (153.69 per cent) and weight gain (3625.40 per cent) both for male and female were observed in 10 gL^{-1} salinity and that did not significantly differed from salinity, 15 gL^{-1}. After 30 days of rearing, similar observations were reported by Darve (2005). She reported that the maximum length gain (33.26 per cent) and weight gain (163.06 per cent) of *O. melastigma* were in 10 gL^{-1} salinity. She had also found that the growth rate by weight in 10 gL^{-1} salinity was significantly different with growth rate in freshwater and 5 gL^{-1} salinities. Takahito and Fujio (1997) reported that the body length of guppy *Poecilia recticulata* reared in seawater was smaller than that of freshwater as control. After 180 days rearing in freshwater they observed that maximum body length of female was 30 mm and for male was 20 mm whereas in sea water female attained maximum body length up to 23 mm and male upto 18 mm. The guppies reared in seawater could propagate in seawater with successful maturation, fertilization and larval development. Totally opposite results were found in the present study where the higher body length of female (36.77 mm) and male (34.27 mm) was observed in seawater. The minimum body length of female (29.87 mm) and male (29.60 mm) was observed in freshwater. But growth and maturation was observed in all salinities. This might be due to the euryhaline nature of this fish species and guppy bears a native of freshwater.

Tsuzuki *et al*. (2000) found that maximum length gain (257.79 per cent) of larvae of *Odontesthes banariensis* in 30 gL^{-1} salinity and for that of juveniles maximum length gain (8.19 per cent) was observed in 0, 5 and 20 gL^{-1} salinities. In other experiment for larvae of *O. hatcheri* they observed maximum length gain (143.82 per cent) in 5 gL^{-1} salinity whereas, for juveniles freshwater is required for maximum length gain (31.93 per cent).

Tsuzuki *et al*. (2000) has reported that in 20 and 30 gL^{-1} salinity larvae of *O. banariensis* attained maximum weight gain (1500 per cent) whereas, 3-4 months juveniles attained maximum weight gain (50 per cent) in 20 gL^{-1} salinity. They also observed that larvae of *O. hatcheri* attained maximum length (1941.67 per cent) in 5 gL^{-1} and juveniles (91.87 per cent) in 5 gL^{-1} salinity. In general, they proved that *O. bonariensis* and *O. hatcheri* were capable of surviving and growing in salinities ranging from freshwater to seawater.

Maximum specific growth rate recorded both for male and female was 4.02 per cent day^{-1} observed in 10 gL^{-1} salinity. Similar results were obtained by Darve (2005). She observed maximum SGR of *O. melastigma* (3.22 per cent day^{-1}) in 10 gL^{-1} salinity after 30 days of rearing. Tsuzuki *et al*. (2000) reported higher SGR of larvae of *O. bonariensis* (9.8 per cent day^{-1}) in 20 and 30 gL^{-1} and *O. hatcheri* (10.6 per cent day^{-1}) in 5 gL^{-1} salinity after four weeks rearing in different salinities. They also reported that juveniles of *O. bonariensis* attained maximum SGR (1.5 per cent day^{-1}) in 20 gL^{-1} salinity whereas, *O. hatcheri* attained maximum SGR (2.3 per cent day^{-1}) in freshwater. Martinez-Palacios *et al*. (2004) reported that the SGR of larvae of *Chirostoma estor estor* was significantly greater at 10 gL^{-1} and 15 gL^{-1} (13.94 per cent day^{-1}) than those seen at 5 gL^{-1} (13.35 per cent day^{-1}) salinity and freshwater (10.86 per cent day^{-1}). These findings are supportive to the present experiment. From the above findings it is clear that intermediate salinity is essential for better growth and survival of euryhaline teleost. Holliday (1969) reported that, the activity levels of euryhaline teleost are often lower in low salinity and energy expenditure is therefore less and the ability to survive and achieve rapid growth rates may thus be increased.

During the present study, cent-per-cent survival was observed in all salinities after 90 days of rearing period. Darve (2005) observed similar results as 97 to 100 per cent larvae of *O. melastigma* were survived in all salinities (0 to 35 gL^{-1}) and survival rate was not significantly differed with each other in all salinities. Johnson and Katavic (1986) observed that survival of larvae of sea bass, *Diacentrachus labrax* consistently increase (90.7 per cent) with reduction in salinity (10 and 20 gL^{-1}) below ambient levels (38 gL^{-1}). At salinities 20, 30 and 38 gL^{-1} mortality increased with elevated temperature *i.e.* survival was greater at 15 °C with 10 gL^{-1} salinity. They observed no beneficial effects of ambient salinity (38 gL^{-1}) on larvae and did not use hyposmotic salinities. Tsuzuki *et al.* (2000) observed cent-per-cent mortality of *O. bonariensis* larvae in freshwater and survival percentage ranged between 57 to 80 per cent in other salinities while 78 per cent survival of larvae of *O. hatcheri* was observed in all salinities after four weeks rearing. Three to four month juveniles of *O. bonariensis* showed 100 per cent survival in all salinity while cent-per-cent mortality was observed only in 30 gL^{-1} salinity. They observed > 95 per cent survival of juveniles of *O. hatcheri* in all salinities (0 to 20 gL^{-1}).

During the present study, better GSI for male recorded was (2.62 per cent) in salinity 10 gL^{-1} and was differed significantly with other salinities except 15 gL^{-1} salinity. Similarly, maximum GSI for female of *O. melastigma* was recorded in salinity 10 gL^{-1} that was not significantly differed with GSI observed in 15 gL^{-1} salinity.

Anonymous (2006) reported that medaka, *Oryzias latipes* male attained maximum GSI (1.3 per cent) and female (11.0 per cent) in 10 ngL^{-1} estradiol and 100 ngL^{-1} estradiol respectively after 21 days rearing. This report also states that after rearing of medaka in trenbolone male attained maximum GSI (1.3 per cent) in 500 ngL^{-1} concentration and female attained maximum GSI (16.0 per cent) in control condition. Bartulovic *et al.* (2006) reported that the average GSI of sand smelt, *Atherina boyeri* males was 2.10 per cent and 1.59 per cent for females. The highest GSI was recorded for males 4.7 per cent and for females 5.4 per cent in April because of high and unpredictable temperature and salinity variation.

Thus, the results obtained during present experiment indicated that 10 to 15 gL^{-1} salinity is optimum salinity range for better survival, maximum growth and early maturation of *O. melastigma.*

References

Anonymous, 2006. *Report of the Initial Work Towards the Validation of the 21 Day Fish Screening Assay for the Detection of Endocrine Active Substances.* Director, Organisation for Economic Co-operation and Development, OECD Environment Health and Safety Publications, Paris, 61 pp.

Bartulovic, V., Glamuzina, B., Conides, A., Gavrilovic, A. and Dulcic, J., 2006. Maturation, reproduction and recruitment of the sand smelt, *Atherina boyeri* Risso, 1810 (Pisces : Atherinidae) in the estuary of Mala Neretva River (southeastern Adriatic, Croatia). *ACTAADRIAT*, 47: 5–11.

Darve, S.I., 2005. Studies on breeding and rearing of estuarine rice fish, *Oryzias melastigma* (McClelland, 1839). *M.F.Sc. Thesis*, Dr. Balasaheb Sawant Konkan Krishi Vidyapeeth, Dapoli, Ratnagiri, India (Unpublished) 52 pp.

Denny, J.S., Spehar, R.L., Mead, K.E. and Yousuff, S.C., 1991. *Guideline for Culturing the Japanese medaka, Oryzias latipes.* U.S. Environmental Protection Agency, Environmental Research Laboratory, Duluth, 38 pp.

Holliday, F.G.T., 1969. The effects of salinity on the eggs and larvae of teleost. In: *Fish Physiology*, W.S. Hoar and D.J. Randall. Academic Press, New York, 1: 293–311.

Johnson, D.W. and Katavic, I., 1986. Survival and growth of Sea bass (*Dicentrachus labrax*) larvae are influenced by temperature, salinity and delayed initial feeding. *Aquaculture*, 52: 11–19.

Kirchen, R.V. and West, W.R., 1976. *The Japanese Medaka: Its Care and Development*. Carolina Biological Supply Company, Burlington, NC, 36 pp.

Manna, A.K., 1989. Behavioural study of weed fish *Oryzias melastigma*. *Environment Ecology*, 7(2): 502–503.

Martinez-Palacios, C.A., Morte, J.C., Tello-Ballinas, J.A., Toledo-Cuevas, M. and Ross, L., 2004. The effects of saline environments on survival and growth of eggs and larvae of *Chirostoma estor estor* Jordan 1880 (Pisces : Atherinidae). *Aquaculture*, 283: 509–522.

Snedecor, G.W. and Cocharn, W.G., 1967. *Statistical Methods*, 6th Edn. Oxford and IBH Publishing Co. Pvt. Ltd., New Delhi, 593 pp.

Takahito, S. and Fugio, Y., 1997. Successful propagation in seawater of the guppy *Poicilia reticulata* with reference to high salinity tolerance at birth. *Fisheries Science*, 63(4): 573–575.

Talwar, P.K. and Jhingran, A.G., 1991. *Inland Fishes of India and Adjacent Countries*. Oxford and IBH Publishing Co. Pvt. Ltd., New Delhi, Bombay, Calcutta, pp. 744–746.

Tsuzuki, M.Y., Aikawa, H., Strussmann, C.A. and Takashima, F.T., 2000. Comparative survival and growth of embryos, larvae, and juveniles of pejerrey *Odontesthes bonariensis* and *O. hatcheri* at different salinities. *J. Appl. Ichthyol.*, 16: 126–130.

Zar, J.H., 2005. *Biostatistical Analysis*, 4th Edn. Pearson education (Singapore) Pvt. Ltd., Indian Branch, Delhi, 663 pp.

Chpater 12

Acute Toxicity of Dimethoate on Freshwater Fish, *Channa gachua*

☆ *R.M. Reddy and V.S. Shembekar*

Introduction

With the onset of green revolution the problem of environmental contamination by excessive use of pesticides can not be overlooked. The unjudicious and ill-programmed use of pesticides has caused serious environmental hazards affecting aquatic fauna. Pesticides enter the water bodies along with agricultural run off water and pollute it (Datta *et al.*, 1992). Pesticides are non-biodegradable and tend to get accumulated in the body of fish and hence get into the body of human through food chain.

In freshwater fauna, fishes are the most important group due to their nutritive value. They are highly sensitive towards changes occurring in their surroundings. Hence, they serve as a biological indicator to detect degree of pollution of water bodies concerned. There is a growing concern about the adverse effects of environmental contamination on aquaculture system (Sarvana *et al.*, 1997).

Toxicology deals with the study of adverse effects of chemicals in biological systems. Exposure to toxic chemicals can produce unexpected effects in nontarget animals (Abdul Naveed *et al.*, 2004; Veronica and Collins, 2003).

The results of toxicity are generally reported in terms of median lethal concentration, LC_{50} or median tolerance limit (Vasait and Patil, 2005). In aquatic toxicology, the traditional LC_{50} test is often used to measure the potential risk of a chemical.

Dimethoate, an organophophate is one of the commonly used pesticides in and around Latur to control pests. Excess use of this mainly affects the non-target organisms, particularly fishes which are highly sensitive to it.

There are many reports about toxicity of pesticide on different fish species. But little information is available on effects of dimethoate on *Channa gachua*. Hence present study was made to evaluate the acute toxicity of dimethoate on *Channa gachua* as well as to evaluate the presumably harmless (safe) concentration of dimethoate.

Materials and Methods

In the present investigation, a freshwater air–breathing fish called *Channa gachua* was selected as test animal and dimethoate–30 per cent EC, was used as a biocide in the bioassay test. Fishes of almost uniform size and weight measuring 12±2 cm and weighing 14±2 gms were selected. These fishes were observed for any pathological symptoms and then placed in a dilute bath (0.1 per cent) of potassium permagnate ($KMnO_4$) for 2 minutes so as to avoid any dermal infection. The fish were then washed with water and acclimatized to laboratory conditions as suggested by Klontz and Smith (1968). The fish were acclimatized for two weeks in 50 litre capacity glass aquaria containing clean, aged and dechlorinated water. The aquaria were properly covered as fish tries to jump out of the aquaria.

The physico-chemical parameters of water were analyzed at regular intervals by following the standard methods suggested by APHA (1998) and IAAB (1998). Every effort was made to provide optimum conditions to the fish. During acclimatization, the fish were provided with a diet consisting of live earthworms. Water was changed everyday. If mortality exceeded 5 per cent in any of the batch of fish during acclimatization, the whole batch was discarded.

For the determination of acute toxicity, the laboratory acclimatized fishes were sorted into 8 batches of 10 each and kept in aquarium with measured quantity of water (1 liter of water per 1 g wet weight biomass).

One day prior to the commencement of experiment, feeding was withdrawn to avoid metabolic differences due to differential feeding. The test fish were also not fed during bioassay test to avoid any change in the toxicity of the pesticides by excretory products.

Stock solution of dimethoate was prepared. For bioassay test, few concentrations from stock solutions were prepared as per dilution technique suggested by APHA (1998).

Preliminary experiments using different concentrations of dimethoate were conducted to find concentration that resulted in 0–100 per cent mortality *i.e.* LC_0 and LC_{100}. After conducting such few initial test range finding experiments *i.e.* pilot reading, suitable dilutions of toxicant were prepared. The range of concentration set for dimethoate was between 5 ppm to 40 ppm (5, 10, 15, 20, 25, 30, 35 and 40 ppm).

For acute toxicity test, the fish were exposed for duration of 24, 48, 72 and 96 hours, to determine the LC_{50} values. Static bioassay studies were carried out as directed by Doudoroff *et al.* (1951). Medium containing pesticide was changed in every 24 hours to maintain the pesticide concentration constant. Fresh stock solutions and required dilutions were prepared for each exposure. A batch of control fish was also run simultaneously along with experimental fish. Dead fish were removed immediately as they deplete dissolved oxygen to a great extent from the aquaria. Every experiment was repeated 5 times at the selected pesticide concentrations, every time noting the number of fish killed at each concentration upto 96 hours and the mean value was taken. These values were used to determine LC_{50} values for 24, 48, 72 and 96 hours. In the present study, LC_{50} values for different intervals of time was determined using 3 methods–Graphical method given by Fisher (1964), Statistical method of Finney (1971) and Dragstedt and Behren's method (1975).

The presumably harmless (safe) concentration of the pesticide was calculated using the formula suggested by Hart, Doudoroff and Greenbank (1945),

$$C = 48 \text{ h } LC_{50} \times \text{application factor } (A)/S^2$$

where,

C: Presumably harmless concentration

S: 24 hr. LC_{50}/48 hr. LC_{50}

A: Application factor, which is 0.3.

The value of application factor was suggested by Henderson *et al.* (1959).

Results

The LC_{50} values for 24, 48, 72 and 96 hours for dimethoate determined by 3 methods and their average value are presented in Table 12.2. By using graphical method the LC_{50} values for 24, 48, 72 and 96 hours was found to be 30.20, 25.12, 23.44 and 19.95 ppm respectively. When statistical method was followed the LC_{50} values obtained for 24, 48, 72 and 96 hours were 27.54, 25.12, 23.99 and 19.50 ppm respectively. After applying Dragstedt and Behren's method (D and B method) for verification, the values were always high as cumulative mortality is calculated in this method. The values of LC_{50} obtained by following this method for 24, 48, 72 and 96 hours were 36.16, 26.70, 24.02 and 21.30 ppm respectively. Comparison of LC_{50} values by different methods showed that the values are close to each other and are not significantly different from one another suggesting that all the three methods are useful in determining LC_{50} values. An average LC_{50} values for 24, 48, 72 and 96 hrs have been 31.29, 25.65, 23.82 and 20.25 ppm respectively. During the period of acute toxicity tests, no mortality was observed in control fish. The results obtained from the acute toxicity of pesticides at different time intervals were utilized for determining the presumably harmless concentration of dimethoate which was found to be 5.17 ppm.

Table 12.1: Physico-chemical Parameters

Parameters	*Value*
Temperature (°C)	28±4
pH	7.4±0.2
Dissolved Oxygen (mg/l)	7.0±1.0
Total hardness $CaCO_3$ (mg/l)	170±8.0

Table 12.2: LC_{50} Values (in ppm) of Dimethoate for the Freshwater Fish, *Channa gachua* Obtained by 3 Different Methods

Sl.No.	*Method Employed*	*LC_{50} Value at*			
		24 hrs	*48 hrs*	*72 hrs*	*96 hrs*
1.	Graphical	30.20	25.12	23.44	19.95
2.	Statistical	27.54	25.12	23.99	19.50
3.	Dragstedt and Behrens	36.16	26.70	24.02	21.30
4.	Average	31.29	25.65	23.82	20.25

Values expressed in ppm are mean of five observations.

Discussion

LC_{50} values have been determined for different fishes in relation to different pesticides by previous workers. Cope (1965) found that these values differ greatly from one animal to another. Such difference in the toxic values could be attributed to various factors such as variability in bioassay technique, difference in size and weight of test fishes.

In the present investigation it is found that percent mortality increases with increase in concentration of pesticides as documented earlier by Oti (2002), Ayuba and Ofojekwu (2005).

From Table 12.2 it is observed that LC_{50} values decreased with increase in exposure period. Similar observations were made earlier by other workers. The toxicity of monochrotophos on the mortality of *Nemacheilus botia*, an edible fish species has been studied by Vasait and Patil (2005). They found LC_{50} as 49.6 and 42.0 ppm for 7 and 14 days exposure period.

Anupama (1987) studied the effect of endosulfan on *Idoplanorbis exustus* exposed to endosulfan for a period of 24, 48, 72 and 96 hours and showed that the LC_{50} values are 0.8, 0.6, 0.4 and 0.3 ppm respectively.

Pankaj Kumar *et al.* (2004) found malathion toxicity on fingerlings and fish of air–breathing cat fish, *Heteropneustes fossilis*. Toxicity for a period of 24, 48, 72 and 96 hours was found to be 19.0, 18.0, 17.5 and 17.0 ppm in developmental stage while later in adult fish it increased to 35.4, 24.0, 35.5 and 33.0 ppm respectively.

Shivalkumar *et al.* (2006) studied acute toxicity of chromium on freshwater fish, *Mystus vitatus*, for a period of 24, 48, 72 and 96 hours and reported LC_{50} value as 82.79, 72.11, 64.42 and 61.67 mg/l respectively.

From the present result and above discussion, it is clear that the LC_{50} values decreased with increase in exposure period suggesting that with increase in duration of exposure the pesticide becomes toxic even at lower concentrations. It is also clear that the acute concentration of dimethoate is harmful to *Channa gachua*. Hence it is recommended that the application of pesticides should be carefully monitored so that concentrations which are harmful to aquatic fauna do not get introduced in the water.

The presumably harmless concentration of dimethoate estimated indicates that any concentration exceeding safe concentration may be hazardous to fish.

References

Anupama, G., 1987. Effects of endosulfan on *Indoplanorbis exustus. M.Phil. Dissertation*, Andhra University, Waltair, 51 pp.

APHA, 1998. *Standard Methods for the Examination of Water and Wastewater*, 20th Edn.. American Public Health Association Washington, DC.

Ayuba, V.O. and Ofojekwu, P.C., 2005. Effects of sublethal concentration of *Datura innoxia* leaf on weight gain in the African catfish, *Clarias gariepinus. Journal of Aquatic Sciences*, 20(2): 113–116.

Bhavan, Sarvana, Zayapragassarazan, P.Z. and Geraldine, P., 1997. Accumulation and elimination of endosulfan and carbaryl in the freshwater prawn, *Macrobrachium malcolmsonii* (H. Milne Edwards). *Poll Res.*, 16 (2) 113–117.

Cope, O. B., Wood, E.M. and Wallen, G.H., 1970. Some chronic effects of 2, 4-D on the blue gills, *Laponis macrochurus. Trans. Am. Fish. Soc.*, 99, 1.

Datta, H.M., Dogra, J.V.V., Singh, N.K., Roy, P.K., Nasar, S.S. Adhikari, T.S., Munshi, J.S.D. and Richmonds, C., 1992. Malathion induced changes in the serum proteins and haematological parameters of an Indian cat fish, *Heteropneustes fossilis* (Bloch). *Bull. Environ, Contam. Toxicol.*, 49: 91–97.

Doudoroff, 1951. Bioassay methods for the evaluation of acute toxicity of industrial wastes to fish. *Sewage Indstr. Waste*, 23: 1397–1380.

Dragstedt and Behren's, 1975. In: *Immunology and Serology*, 3rd Edn., (Ed.) P.I. Carpenter. W.B. Saunders Co., Philadelphia, p. 286.

Finney, D.J., 1971. *Probit Analysis*, 3rd Edn. Cambridge University, Press, London, p. 25–66.

Fisher, R.A., 1964. *Statistical Methods for Research Workers*. Oliver and Body, London.

Hart, W.B.P., Doudoroff and Greenbank, J., 1945. The evalution of the toxicity of industrial wastes, chemicals and other substances to freshwater fishes. *Atlantic Refining Co.*, p. 317.

Henderson, C., Pickering, Q.H. and Tarzwell, C.M., 1959. The relative toxicity of ten chlorinated insecticides to four species of fish.*Trans. Amer. Fish Soc.*, 88: 23–37.

IAAB, 1998. *Methodology of Water Analysis*, 2nd Edn. Indian Association of Aquatic Biologists, Hyderabad.

Klontz, G.W. and Smith, L.S., 1968. Methods of using fish as biological research subject. In: *Methods of Animal Experimentation III*, (Ed.) W.I. Gay. Academic Press New York, p. 322–385.

Kumar, Pankaj, Sharma, B. and Mishra, A.P., 2004. Efficacy of malathion on mortality of a freshwater air–breathing cat fish, *Heteropneustes fossilis* (Bloch) during different developmental stage. *Ecol. Env. and Cons.* 10 (1): 47–52.

Naveed, Abdul, Venkateshwarlu, P. and Janaiah, C., 2004. The action of sub lethal concentration of endosulfan and kelthane on regulation of protection metabolism in the fish, *Clarias batrachus. Nat. Environ. and Pollution. Tech.*, 3(4): 539–544.

Oti, E.E., 2002. Acute toxicity of cassava mill effluent to the African catfish fingerlings. *Journal of Aquatic Sciences*, 17 (1): 31–35.

Shivalkumar, R., 2005. Endosulfan induced metabolic alteration in freshwater fish, *Catla catla. Ph.D. Thesis.* Karnataka University, Dharwad, Karnataka, India.

Vasait, J.D. and Patil, V.T., 2005. The toxic evaluation of organophosphorous insecticide monocrotophos on the edible fish species, *Nemacheilus botia. Ecol. Env. and Cons.*, 8(1): 107–109.

Veronica, W. and Collins, P.A., 2003. Effect of cypermethrin on the freshwater crab, *Trichodactylus. Bull. Environ Contam. Toxicol.*, 71(1): 106–113.

Chapter 13

Impact of Copper Sulphate on the Oxygen Consumption in the Freshwater Fish, *Catla catla*

☆ *C.M. Bharambe*

Introduction

Level of Heavy metals in aquatic ecosystem and extensive uses of various chemical contaminantes are known to adversely affect growth, survival and physiology of freshwater fauna (Sarojini and Indira, 1990). The toxicity of the chemical contents in vertebrates is mainly attributed to the central nervous system (Nagbhushanam and Diwan, 1972). Respiration is the sign of life and index of all biochemical activities that occur due in the effect of toxicants on the overall metabolism of the exposed animals (Zeuthen, 1970 and Akerlund, 1974). The changes in the oxygen consumption due to pollution stress create a physiological imbalance in the organisms. The gills are the major organs in the respiratory process of aquatic animals like gastropod including snails (Sokolova and Portner, 2003). When the toxic contaminants are water born, the gills are the sites for damage, which can be easily assayed (Mali and Ambhore, 2003 and Sokolova Portner, 2003).

The literature of many authors reveals that considerable work has been done on the effect of pollutants on the respiration of marine animals but comparatively little attention has been focused on their freshwater counterparts particularly Nalganga reservoir fish. Hence the thought was undertaken to study respiratory metabolism of the freshwater fish, *Catla catla* after exposure to copper sulphate for varying period of time such as 24, 48, 72 and 96 hours.

Materials and Methods

A freshwater fish *Catla catla* were collected from the Nalganga Reservoir, Nalgangapur, Buldana District and were brought to the laboratory. Healthy adult fish of uniform weight (30 to 40gm) were sorted, cleaned, maintained separately in the groups of five in plastic troughs. They were acclimated to the laboratory conditions for 5 days prior to experiment.

The animals were fed with small pieces of goat muscle to avoid effects of starvation. The animals were subjected to sublethal concentration of $CuSO_4$ and oxygen consumption was studied after 0, 24, 48, 72 and 96 hours of exposure.

The respiratory metabolism was studied by modified Winkler's method (Welsh and Smith, 1959). The set designed by Bharambe (2001) was used to determine the oxygen consumption. The animal was allowed to stay in the chamber for one hour and at the end, the final sample was collected. Oxygen consumption in initial and final water samples was determined for one hour by standard method suggested (APHA, 1989).

Results and Discussion

The freshwater fish *Catla catla* showed variations in total oxygen consumption and rate of oxygen consumption when exposed to copper sulphate up to 96 hours. Investigation was stated that gradual-decreasing trend was seen in total oxygen consumption and rate of oxygen consumption up to 96 hours as compared to control (Table 13.1).

Table 13.1: Effect of Copper Sulphate on Total Oxygen Consumption and Rate of Oxygen Consumption in Fish *Catla catla*

Sl.No.	*Duration of Exposure Period (hrs)*	*Average Weight of Fish (gm)*	*Total Oxygen Consumption in Fish (mg of O_2/animal/hr)*	*Rate of Oxygen Consumption in Fish (mg of O_2/gm/hr)*
1.	00 (control)	34	2.982±0.45	0.0877±0.02
2.	24	32	1.688±0.22	0.0528±0.01
3.	48	35	0.965±0.25	0.0276±0.07
4.	72	38	0.885±0.18	0.0233±0.08
5.	96	40	0.756±0.20	0.0189±0.04

Values are the mean of five replicates±SD.

The results obtained indicate that the total oxygen consumption of fish *Catla catla* decreased gradually when exposed with sublethal concentration *i.e.* 2 ppm of copper sulphate. The control set showed maximum respiratory metabolism, but when the animals were exposed to copper sulphate, they showed marked decline in total oxygen consumption and rate of oxygen consumption.

It has been reported that oxygen consumption represents the physiological state of metabolic activity and may be an indictor of metabolic stress. The pollutants may induce stress to exposed animals (McMahon and Russell, 1981 and Prosser, 1991). Many workers have shown the harmful effects of heavy metals on histological structure of gills of crustaceans (Dixcon and Leduc, 1981; Sarojini and Indra, 1990; Nilknath and Sawant, 1993; Ramanna Rao and Ramamurthy 1996; Khan *et al.*, 2000). The decline in the rate of oxygen consumption may be the result of formation of coagulated mucous over the gills and body surface of the fish. Similar changes were reported (Chinnayya, 1971; Nagabhushanam and Diwan, 1972 and Nagbhushanam and Kulkarni, 1981). A significant drop in rate of oxygen consumption in *Cyprinus carpio* exposed to both fenvalerate and cyper methrin was observed (Malla Reddy, 1987). The oxygen consumption was reduced in all the tissues in an edible freshwater field crab, *Barytelphusa guerini* studied (Reddy *et al.*, 1993) when exposed to cadmium chloride. Studies on respiratory metabolism in the freshwater bivalve, *Corbicula striatllea* exposed to carbaryl and cypermethrin was reported by Hemmingsen, 1960; Marsden, 1979; Ivleva, 1980; Debbagh and Marina, 1986. They found a decrease in oxygen consumption with the increase in the exposure period. Asifa and Vasantha, (2001) studied effect of ammonia on respiratory activity of air breathing fish *C. batrachus* and showed that the acute and subacute treatment on animal lead to severe disruption of oxidation reduction process and suppressed tissue respiration. Tilak and Satyavardhan (2002) showed that the amount of oxygen consumption was initially increased and then gradually decreased when the fish, *C. punctatus* were treated with fenvalerate.

The inhibition in oxygen consumption may be due to disintegration or rupture of respiratory epithelium and coagulation of mucus film over the gill surface (Jones 1947; Pones, 1947; Chinnayya, 1971). As a result, the absorption of oxygen by the gills from the external milieu might have adversely affected.

References

Akerlund, G., 1974. Oxygen consumption in relation to environmental oxygen concentrations in the ampullarid snail, *Marisa cornuarietis* (L). *Comp. Biochem. Physiol.*, 47A: 1065–1975.

Mali, R.P. and Ambore, N.E., 2003. Impact of copper sulphate on the oxygen consumption in the freshwater female Crab. *J. Comp. Toxicol. Physiol.*, 1(1): 14–18.

APHA, 1989. *Standard Methods for the Examination of Water and Wastewater,* 17th edn. American Public Health Association, Washington, DC.

Asifa, P. and Vasantha, N., 2001. Effect of ammonia on respiratory activity of an air breathing fish, *Clarius batrachus. Nat. Conference. Hyderabad* Abs. No. 59.

Bharambe, C.M., 2007. Metabolic responses of Snail *Bellamya bengalensis* to changing temperature, pH, salinity and day time. *M.Phil. Disser.*, SGB Amt. Uni. Amt., p. 17–23.

Chinnayya, B. and Kulkarni, G.K., 1971. Effect of heavy metals on the oxygen consumption by the shrimp, *Carolina rajadhari. Ind. J. Expt. Biol.*, 9(2): 277–278.

Debbagh, K.Y. and Marina, B.A., 1986. Relationship between oxygen uptake and temperature in the terrestrial isopod *Porcellionides pruinosus. J. Arid Environ.*, 11: 227–233.

Dixon, G.D. and Leduc, G., 1981. Chronic cyanide poisoning of rainbow trout and its effects on growth, respiration and Imer histopathology. *Arch. Enviro. Conta. and Toxicol.*, 10: 117–131.

Hemmingsen, A.M., 1960. Energy metabolism as related to body size and respiratory surfaces, and its evolution. *Rep. Steno. Mem. Hosp. Nordisk. Insulin Lab.*, p. 9: 1–110.

Ivleva, I.V., 1980. The dependence of crustacean respiration rate on body mass and habitual temperature. *Int. Rev. Gas. Hydrobiology*, 65: 1–47.

Jones, J.R.E., 1947. The oxygen consumption of *Gasterostes acculeatus* in toxic solutions. *J. Expt. Biology*, 23: 291–311.

Khan, A.K., Patel, R.T. and Shaikh, F.I., 2000. Effect of mercuric chloride on the gills of the freshwater crab, *Barytelphusa guerini. International Conference on Probing in Biological Systems*, Mumbai Abstract, pp. 134.

Malla Reddy, P., 1987. Effect of fenvalerate and cypermethrin on the oxygen consumption of a fish, *Cyprinus carpio. Mendel*, 4: 209–211.

Marsden, I.D., 1979. Seasonal oxygen consumption of the salt marsh isopod. *Sphaeroma regicauda. Mar. Biol.*, 51: 329–337.

McMahon, F.R. and Russell, W.D., 1981. The effects of physical variables and acclimation on survival and oxygen consumption in the high Littoral Salt-Marsh snail, *Melampus bidentatus* Say. *Biol. Bull.*, 161: 246–269.

Nagbhushanam, R. and Kulkarni, G.K., 1981. Freshwater palaemonid prawn, *Macrobrachium kistnensis*: Effect of heavy metal pollutants. *Proc. Ind. Sci. Acad.*, B.47(3): 380–386.

Nagbhushanam, R. and Diwan, A.D., 1972. Effect of toxic substance on oxygen consumption of the freshwater crab, *Barytelphusa cunicularis. Nat. Sci. J.*, 11: 127–129.

Nilkanth, G.V. and Savant, K.B., 1993. Studies on accumulation and histopathology of gills after exposure to sublethal concentration of hexavalent chromium and effect on oxygen consumption in *Scylla serrata. Poll. Res.*, 12(1): 11–18.

Prosser, C.L., 1991. *Environmental and Metabolic Animal Physiology*. Wiley-Liss, New York.

Ramanna, Rao and Ramamurthy, M.V., 1996. Histopathologieal effects of sublethal mercury on the gills of freshwater filed crab, *Oziotelphusa senex* (Fabricus) *Ind. J. Comp. Ani. Physiol.*, 14(2): 33–38.

Reddy, S.L.N. and Venugopal, N.B.R.K., 1993. Effect of cadmium on oxygen consumption in freshwater field crab, *Barytelphusa guerini. J. Environ. Biol.*, 14(3): 203–210.

Sarojini, R. and Indira, B., 1990. Lethal and sublethal effects of antifouling organo metallic compounds on the gills of *Caridina webberi. Himalay. J. Enviro. Zool.*, 4(1): 40–45.

Sokolova, I.M. and Portner, H.O., 2003. Seasonal respiration in the marsh periwinkle, *Littoring irroata. Biol. Bull.*, 154: 322–334.

Sokolova, I.M. and Portner, H.O., 2003. Metabolic plasticity and critical temperatures for aerobic scope in a eurythermal marine invertebrate (*Littorina saxatilis*, Gastropoda : Littorinidae) from different latitudes. *J. of Exp. Biol.*, 206: 195–207.

Tilak, K.S. and Satyavardhan, K., 2002. Effect to fenvalerate on the oxygen consumption and hematological parameters in the fish, *Channa puntatus* (Bloch.) *J. Aqua. Biol.*, 17(2): 81–86.

Welsh, J.H. and Smith, R.I., 1959. *The Laboratory Exercise in Invertebrate Physiology*. Burgess, Publication, Company, U.S.A., Minneapolis, p. 10–53.

Zeuthen, E., 1970. Rate of living as related to body size in organism. *Pal. Arch. Hydrobiol.*, 17: 21–30.

Chapter 14

Effect of Temperature on Biochemical Composition of *Lamellidens marginalis*

☆ *S.S. Surwase, D.A. Kulkarni and R.S. Chati*

Introduction

Changes in biochemical constituents are pronounced in invertebrates which are cyclic in reproduction since a great amount of energy must be channelised to the gonad during reproduction. This is reflected in the deposition or depletion of the nutrients with the advent or departure of the reproductive period (Lambert or Dahnel, 1974). The aspect of energy metabolism and reproduction has been reported for a number of species of bivalves due to their commercial importance and edibility value. But the relative influence of gonad development on the distribution and storage of biochemical constituents in different body parts has been examined in only a few cases (Giese, 1969), Gabbolt (1975,1976), Bayne (1976 a) and DaZwaan (1983) have reviewed much of the work on biochemical changes, particularly the carbohydrates. A review of lipids in marine invertebrates including bivalves is given by Giese (1966), Lawrence (1976) and Voogt (1983). Seasonal variations in the biochemical composition of *Mytilus edulis* in British waters have been reported by Daniell (1920, 1921, 1922), Williams (1969) and Bayne and Thompson (1970). Ansell *et al*. (1964) determined seasonal changes in biochemical composition of adductor muscle, mantle, siphon, visceral mass (gonad), digestive gland and foot in *Mercenaria mercenaria*.

The amount of nutrients mobilized for the gonads could be affected by the energy requirements of somatic growth and basic metabolism. These relationships may change according to age and reproductive stage and also according to changes in food concentration and temperature conditions in the environment (Sastry, 1979).

The biochemical constituents shown cyclic changes in reproduction due to great amount of energy to be channalized to the gonad during reproduction (Muley, 1988).

Hence in the present study an attempt is made by effect of rise in temperature to the *Lamellidens marginalis* on biochemical composition.

Materials and Methods

The biochemical analysis from different soft tissues *viz.* mantle, hepatopancreas and gonad was done from the animals belonging to respective controls and experimental groups of rise in temperature on 2nd, 6th and 12th day. The biochemical constituents in different seasons of freshly collected animals in December 2001 (Winter), May 2002 (Summer) and July 2001 (Monsoon). Every time samples were pooled from 5 different animals from each group to estimate total protein (according to Lawry *et al.*, 1951), glycogen (according to Dezwaan and Zandee, 1972), lipid (Barnes and Blackstock, 1973). All the estimations were done from the respective groups within a day in each season after pooling the samples. Triplicate values of each biochemical constituents were subjected for statistical confirmation using student's "t" test (Dowdeswell, 1957). Percentage differences were also calculated between each control and experimental groups in every season.

Results and Discussion

Effect of rise in different temperature in different season on the biochemical constituents of mantle, hepatopancreas and gonad tissue of *Lamellidens marginalis* are expressed in Tables 14.1, 14.2 and 14.3. In the present investigation the biochemical composition of controls groups in mantle the glycogen content did not change significantly in winter and summer but significantly decreased in monsoon ($P < 0.05$) and in winter ($P < 0.01$). The protein content non significantly decreased from winter to summer.

In hepatopancreas the glycogen content did not change significantly in winter and summer but significantly increased in monsoon ($P < 0.01$) than summer. There was no significant change in the content from monsoon to winter. The protein content significantly decreased from winter to summer ($P < 0.001$) but significantly increased in monsoon ($P < 0.01$). There was no significant change in the content from monsoon to winter. The lipid content significantly increased from winter to summer ($P < 0.001$) but from summer to monsoon there was no significant change in the content. From monsoon to winter the content significantly decreased ($P < 0.001$). Thus it is found that the glycogen and protein contents were high in monsoon and winter and lipid in summer and monsoon.

In gonad the glycogen content significantly decreased from winter to summer ($P < 0.5$) but significantly increased in monsoon ($P < 0.01$). The content did not change significantly from monsoon to winter. The protein content did not change significantly decreased in monsoon ($P < 0.01$). The content significantly increased from monsoon to winter ($P < 0.001$). The lipid content did not change significantly from winter to summer but significantly increased in monsoon ($P < 0.05$). The content significantly decreased from monsoon to winter ($P < 0.01$). Thus it is found that the glycogen content was high in monsoon and winter, protein in winter and summer and lipid in monsoon.

In experimental groups the results of the present investigation are compared with controls of the respective days 2nd, 6th and 12th day at different temperature in different seasons in winter at 28°C the biochemical composition in mantle the glycogen content did not show significant change on 2nd and 6th day but significantly increased on 12th day ($P < 0.001$) (5.87 per cent). The protein content also did not show significant change on 2nd and 6th day but on 12th day it significantly decreased ($P < 0.001$) (20.0 per cent). The lipid content did not show significantly change on 2nd and 12th day but on 6th day it significantly increased ($P < 0.001$) (19.0 per cent).

Table 14.1: Effect of Rise in Temperature on the Biochemical Constituents of Mantle Tissue of *Lamellidens marginalis* in Three Different Seasons

Season	*Biochemical Constituents*	*2nd Day*		*6th Day*		*12th Day*	
		Control	*28°C*	*Control*	*28°C*	*Control*	*28°C*
Winter	Protein	6.3121±0.4313	8.1321±0.1214 (28.83)	7.1349±0.1314	6.1389±0.1218 (13.95)	8.1341±0.1349	4.1312±0.1318 (49.21)
	Glycogen	8.1321±0.1423	12.1341±0.3412 (49.21)	10.1421±0.3412	7.1342±0.2138 (29.65)	10.1412±0.1382	12.4131±0.1213 (22.40)
	Lipid	5.2132±0.176	6.1314±0.1314 (17.61)	4.1341±0.1389	6.6123±0.1218 (59.94)	7.1404±0.1381	4.5123±0.1214 (36.80)
Season	*Biochemical Constituents*	*2nd Day*		*6th Day*		*12th Day*	
		Control	*32°C*	*Control*	*32°C*	*Control*	*32°C*
Summer	Protein	3.1218±0.1562	4.4541±02155 (42.22)	4.2171±0.2133	4.1349±0.1238 (1.94)	4.3891±0.3421	3.4121±0.1541 (22.25)
	Glycogen	14.1213±0.1368	9.7992±0.0968 (30.60)	11.1421±0.4312	9.1821±0.4134 (17.59)	11.2134±0.2314	13.5012±0.1318 (20.40)
	Lipid	7.1441±0.0912	9.2195±0.0905 (29.05)	7.6004±0.1567	9.2612±0.4312 (21.85)	7.4312±0.1608	10.1234±0.1345 (36.22)
Season	*Biochemical Constituents*	*2nd Day*		*6th Day*		*12th Day*	
		Control	*30°C*	*Control*	*30°C*	*Control*	*30°C*
Monsoon	Protein	4.1212±0.5112	4.3321±0.1238 (5.11)	4.1342±0.1231	4.7341±0.8123 (2.4)	4.3421±0.1342	4.1342±0.4213 (4.78)
	Glycogen	9.2321±0.264	9.1214±0.1234 (1.19)	9.1242±0.1234	9.1341±0.1341 (1.085)	8.1342±0.1346	8.1451±0.2142 (0.1340)
	Lipid	5.892±0.3803	6.1891±0.3412 (5.04)	4.1342±0.1315	4.1314±0.1381 (0.6)	5.1312±0.1341	4.1238±0.1346 (19.63)

All values are expressed in mg/100gm.

Table 14.2: Effect of Rise in Temperature on the Biochemical Constituents of Hepatopancreas Tissue of *Lamellidens marginalis* in Three Different Seasons

Season	*Biochemical Constituents*	*2nd Day*		*6th Day*		*12th Day*	
		Control	*28°C*	*Control*	*28°C*	*Control*	*28°C*
Winter	Protein	10.2124±0.1342	11.3634±1.1489 (11.27)	4.1214±0.1245	6.1214±0.1218 (48.52)	10.1101±0.5412	5.1213±0.1242 (49.34)
	Glycogen	13.1412±0.1218	9.1234±0.1343 (30.57)	8.1428±0.1423	9.1121±0.1441 (11.90)	9.1421±0.2412	10.1213±0.1245 (10.71)
	Lipid	8.1412±0.1215	9.1214±0.1248 (12.03)	6.1214±0.1213	9.4121±0.4502 (53.75)	6.268±0.2421	5.421±0.2134 (13.65)
Season	*Biochemical Constituents*	*2nd Day*		*6th Day*		*12th Day*	
		Control	*32°C*	*Control*	*32°C*	*Control*	*32°C*
Summer	Protein	4.1214±0.2412	4.1232±0.1242 (0.9)	4.3424±0.07421	4.1341±0.2121 (4.79)	3.9912±0.4123	6.1212±0.8142 (53.36)
	Glycogen	8.1214±0.1214	8.5421±0.1021 (34.41)	8.4213±0.2129	10.1242±0.7142 (-20.22)	9.1242±0.2421	10.1242±0.4213 (10.95)
	Lipid	11.1212±0.2413	10.1241±0.2134 (8.96)	12.4230±0.2912	11.4222±0.3421 (96.61)	9.1421±0.2120	10.8334±0.1234 (18.46)
Season	*Biochemical Constituents*	*2nd Day*		*6th Day*		*12th Day*	
		Control	*30°C*	*Control*	*30°C*	*Control*	*30°C*
Monsoon	Protein	10.212±0.1245	7.1418±0.1417 (30.06)	8.5279±0.08897	6.4135±0.2134 (24.79)	7.8821±0.1249	8.0212±0.1247 (-1.76)
	Glycogen	9.4212±0.1241	11.2134±0.1248 (19.02)	9.4221±0.1421	9.4121±0.4213 (0.106)	10.7256±0.1246	10.4567±0.1245 (2.50)
	Lipid	11.1242±0.3422	11.1281±0.241 (0.35)	9.6218±0.1248	4.214±0.1242 (95.62)	10.2134±0.4213	10.4620±0.1267 (-2.43)

All values are expressed in mg/100gm.

Table 14.3: Effect of Rise in Temperature on the Biochemical Constituents of Gonad Tissue of *Lamellidens marginalis* in Three Different Seasons

Season	*Biochemical Constituents*	*2nd Day*		*6th Day*		*12th Day*	
		Control	*28°C*	*Control*	*28°C*	*Control*	*28°C*
Winter	Protein	13.2248±0.2126	13.4246±0.2435 (1.51)	3.4321±0.1448	7.4233±0.4420 (36.65)	8.9929±0.2125	6.4234±0.1246 (28.57)
	Glycogen	11.1243±0.1284	10.1424±0.4213 (77.99)	8.4342±0.2422	10.1224±0.4235 (94.98)	12.1423±0.1048	11.1234±0.1448 (8.39)
	Lipid	11.1243±0.2474	11.2721±0.1348 (1.32)	16.2622±0.4321	10.1224±0.1438 (37.75)	10.1048±0.2123	14.5632±0.1928 (44.12)
Season	*Biochemical Constituents*	*2nd Day*		*6th Day*		*12th Day*	
		Control	*32°C*	*Control*	*32°C*	*Control*	*32°C*
Summer	Protein	5.2134±0.4348	4.9348±0.2461 (5.34)	4.3242±0.2662	4.4623±0.1296 (3.19)	10.2432±0.2464	6.2212±0.3246 (39.26)
	Glycogen	9.2241±0.2460	10.2224±0.2322 (10.82)	10.4624±0.1432	9.2234±0.2134 (11.84)	10.1242±0.2462	10.4624±0.2802 (3.34)
	Lipid	13.2126±0.4321	15.2324±0.2434 (15.28)	16.3246±01.21	10.2434±0.3245 (37.25)	14.3436±0.6243	13.2124±0.4134 (7.88)
Season	*Biochemical Constituents*	*2nd Day*		*6th Day*		*12th Day*	
		Control	*30°C*	*Control*	*30°C*	*Control*	*30°C*
Monsoon	Protein	7.1434±0.2434	6.1426±0.1241 (14.01)	10.3862±0.2436	9.2668±0.2341 (10.77)	10.6212±0.3298	10.2438±0.2340 (3.55)
	Glycogen	10.1221±0.1346	12.4662±0.4126 (-23.15)	10.2242±0.2124	9.2428±0.2462 (9.59)	10.8239±0.3421	10.1821±0.2462 (5.92)
	Lipid	14.2421±0.2146	15.1235±0.4331 (6.18)	10.2242±0.1238	10.1223±0.2362 (0.99)	12.2602±0.3341	12.2214±0.6512 (0.36)

All values are expressed in mg/100gm.

In hepatopancreas the glycogen content did not show significantly change on 2nd and 6th day but on 12th day it significantly increased ($P < 0.01$) (16.2 per cent). The protein content significantly increased on 2nd and 6th day ($P < 0.05$) (37.11 and 22.6 per cent, respectively) but on 12th day it significantly decreased ($P < 0.01$) (33.3 per cent). The lipid content did not show significantly change on 2nd and 12th day but on 6th day it significantly increased ($P < 0.01$) (16.3 per cent).

In gonad the glycogen content did not show significantly change on 2nd and 6th day but significantly decreased on 12th day ($P < 0.001$) (5.8 per cent). The protein content did not show significantly change on 2nd and 12th day but significantly decreased 6th day ($P < 0.001$) (41.8 per cent). The lipid content significantly increased on 2nd and 12th day ($P < 0.01$) (40.2 and 13.3 per cent respectively) but on 6th day it significantly decreased ($P < 0.001$) (20.3 per cent).

In summer at 32°C the biochemical composition in mantle the glycogen content significantly decreased on 2nd ($P < 0.001$) (11.24 per cent) but on 6th and 12th day there was no significantly change in the content. The protein content significantly increased in 2nd and 12th day ($P < 0.01$) (97.9 and 26.5 per cent respectively) but on 6th day it significantly decreased ($P < 0.01$) (33.7 per cent). The lipid content decreased significantly on 2nd day ($P < 0.001$) (5.3 per cent) but on 6th day and 12th day it significantly increased ($P < 0.01$) (40.5 and 10.4 per cent) respectively. In hepatopancreas the glycogen content did not show significantly change on 2nd and 6th day but on 12th day it significantly decreased ($P < 0.05$) (4.0 per cent). The protein content significantly increased on 2nd ($P < 0.01$) and 12th ($P < 0.05$) day (48.5 and 27.3 per cent, respectively), but on 6th day it significantly decreased ($P < 0.05$) (17.9 per cent). The lipid content significantly decreased on 2nd day ($P < 0.01$) (11.3 per cent) and increased on 12th day ($P < 0.01$) (8.6 per cent) but on 6th day there was no significantly change in the content.

In gonad the glycogen content significantly decreased on 2nd and 12th day ($P < 0.01$) (6.3 and 13.17 per cent respectively), but on 6th day it significantly increased ($P < 0.05$) (2.1 per cent). The protein content did not show significantly change on 2nd and 6th day but on 12th day it significantly decreased ($P < 0.01$) (15.3 per cent). The lipid content significantly increased on 2nd ($P < 0.05$) and 12th ($P < 0.01$) day (8.8 and 9.6 per cent respectively) but on 6th day it significantly decreased ($P < 0.01$) (8.3 per cent)

In monsoon at 30°C the biochemical composition in mantle the glycogen content significantly decreased on 2nd and 12th day ($P < 0.01$) (11.9 and 7.5 per cent respectively) but on 6th day the content significantly increased ($P < 0.01$) (4.2 per cent). The protein content significantly decreased on 6th ($P < 0.05$) and 12th ($P < 0.01$) day (6.6 and 19.6 per cent respectively) but on 2nd day there was no significant change in the content. The lipid content significantly increased on 2nd day ($P < 0.05$) (15.9 per cent) but significantly decreased on 12th day ($P < 0.01$) (15.4 per cent). There was not significant change in the content on 6th day.

In hepatopancreas the glycogen content did not show significant change on 2nd, 6th and 12th day. The protein content significantly decreased on 2nd day ($P < 0.001$) (30.06 per cent) but there was no significant change on 6th and 12th day. The lipid content significantly decreased on 6th ($P < 0.001$) (95.62 per cent) but did not show significant change on 2nd and 12th day.

In gonad the glycogen content significantly decreased on 2nd ($P < 0.01$) and 6th ($P < 0.05$) day (23.15 and 9.59 per cent respectively), but on 12th day there was no significant change in the content. The protein content significant change of the content. The protein content significantly decreased on 2nd ($P < 0.01$) and 12th ($P < 0.01$) day (14.01 and 3.55 per cent respectively) but on 12th day there was no significant change in the content. The lipid content significantly increased on 2nd and 6th day ($P < 0.05$) (6.18 and 10.99 per cent respectively) but on 12th day there was no significant change in the content.

In the present study in *L. marginalis* has been observed that higher the temperature more metabolic shifts in glycogen, protein and lipid occur as the experimental temperature increase at 32°C in summer but in winter 28°C temperature caused more effects on metabolic shifts in the biochemical constituents than at the same temperature in monsoon. It has been observed that summer in aquaria at 32°C the animal showed wide opening of the shell valve and more extension of foot. This caused decrease in the metabolites of foot and 2nd, 6th and 12th day; particularly glycogen and lipid. It is also observed that 32°C the growth of the gametes from gonad enhanced and perhaps due to heavy energy demand at this temperature for maintenance and gamete growth that the glycogen content decreased on 2nd day, while lipid on 6th and glycogen and protein on 12th day. In monsoon the rise in temperature caused increase in the respiration on 12th day it is likely that the metabolites from mantle, protein from gonad are more utilized. In this season also gametes, development enhanced due to rise in temperature to 30°C and the lipid content increased in the gonad. Similar trends are observed by Vedapathak *et. al* 1987, Widdows and Bayne 1971, Gabbott and Bayne 1973, Bayne and Thompson 1970, Rise in temperature increase metabolism in Bivalves is well known for Indian species Mane 1975, Mane and Talikhedkar 1976.

In *Lamellidens marginalis* steroidogenesis in different soft tissues was reported due to pesticide toxicity by Swami *et al.* (1983). Tissues steroidogenesis in different tissues of *L. marginalis* as observed in the present study was probably oriented towards the formation of corticosteroids since stress conditions elevate the corticosteroids in the blood of the animals. The possibility of the tissue metabolism to synthesizing, antinflammatory and anti-alfergic cortisone derivatives from the cholesterol and other precursor need elucidation. Mayer and Meyer (1971) stated that since the bio-synthesis of fatty acids and steroids requires a large portion of the cells energy it would evidently be of considerable selective advantage to the organism to abandon these pathway.

Conclusion

Gonad growth and gametogenesis found to be dependent upon direct intake of food material during monsoon at which time proteins and lipid contents increased in gonad and hepatopancreas. Due to rise in temperature 32°C during summer, more metabolic shifts in glycogen, protein and lipids occur. In winter rise in temperature in 28°C caused more effects on metabolic shifts than at the same temperature in monsoon. Due to exposure to atmospheric air more changes in glycogen, protein and lipid occurred during summer than in monsoon and winter. However, mantle tissue was more affected in all the seasons due to changes in temperature. In summer due to exposure atmospheric air the content decreased from mantle, hepatopancreas and gonad in summer season. Due to exposure to atmospheric air the content increased from gonad and hepatopancreas.

Acknowledgement

Authors wish to acknowledge the Principal and Head, Shikshan Maharshi Dnyandev Mohekar College, Kallam for providing necessary facilities during this work.

References

Ansell, A.D., Loosmore, F.A. and Lander, K.F., 1964. Studies on the hard-shell clam, *Venus mercenaria* in British Waters. II. Seasonal cycle in condition and biochemical composition. *J. Appl. Ecol.*, 1: 83–95.

Barnes, H. and Blackstock, J., 1973. Estimation of lipids in marine animals and tissues: Detailed investigation of the sulphophosphovanillin method for total lipids. *J. Expt. Mar. Biol. Ecol.*, 12(1): 103–188.

Bayne, B.L., 1976. *Marine Musels: Their Ecology and Physiology*. Cambridge University Press, Cambridge, London, New York, Melaborne, p. 1–495.

Bayne, B.L. and Thompson, R.J., 1970. Some physiological consequences of keeping *Mytilus edulis* in the laboratory. *Hegol. Wiss. Meersuutes*, 12: 526–552.

Daniell, R.J., 1920. Seasonal changes in the biochemical composition of the mussel (*Mytilus edulis*). *Rep. Lancashire, Sea. Fish. Lab.*, p. 74–84.

Daniell, R.J., 1921. Seasonal changes in the chemical composition of the mussel (*Mytilus edulis*). *Ibid*, p. 205–221.

Daniell, R.J., 1922. Seasonal changes in the chemical composition of the mussel (*Mytilus edulis*) concluded. *Ibid.*, p. 27–50.

Dezween, A., 1983. Carbohydrate catabolism in bivalves. In: *The Mollusca*, (Ed.) K.M. Wilbur. Academic Press, New York, London, 1: 137–175.

Dezween, A. and Zandee, D.I., 1972. Body distribution and seasonal changes in glycogen content of the common sea mussel, *Mytilus edulis. Comp. Biochem. Physiol.*, 43(A): 53–58.

Dowdeswell, H., 1957. *Practical Animal Ecology*. Methum and Co. Ltd., London.

Gises, A.C., 1969. A new approach to the biochemical composition of the molluscan body. *Oceanogr. Mar. Biol.*, 7: 175–229.

Gabbott, P.A., 1976. Energy metabolism. In: *Marine Mussels*, (Ed.) B.D. Bayne. Cambridge University Press, London, New York.

Gabbott, P.A. and Bayne, B.L., 1973. Biochemical effects of temperature and nutritive stres on *Mytilus edulis* L. *J. Mar. Biol. Assoc.*, U.K., 53: 269–286.

Lambert, P. and Dahnel, P.A., 1974. Seasonal variations in biochemical composition during the reproductive cycle of the intertidal gastropod, *Thais lamellosa* (Gastropoda, Prosobranchaia). *Can. J. Zool.*, 52: 305–318.

Lowry, O.H., Resenburough, N.I., Farr, A.L. and Randall, R.J., 1951. Protein measurements with Folin phenol reagent. *J. Chem.*, 193: 265–175.

Mane, U.H. and Talikhedkar, P.M., 1976. Respiration of the wedge clam, *Donax cuneatus. Indian J. Mar. Sci.*, 5: 243–246.

Mane, U.H., 1975. Oxygen consumption of the clam, *Katelysia opima* in relation to environmental conditions. *Broteria*, 60(1–2): 33–38.

Meyer, F. and Meyer, H., 1971. Loss of fattyacid bio-synthesis in flatworms. In: *Comparative Biochemistry of Parasites*, (Ed.) H. Vanden Bossche. Academic Press, New York, p. 383–393.

Muley, S.D., 1988. Reproductive physiology of Lamellibranch molluscs from Maharashtra state. *Ph.D. Thesis*, Marathwada University, p. 1–292.

Sastry, A.N., 1979. Petecypoda (excluding ostreidea). In: *Reproduction of Marine Invertebrates*, (Eds.) A.C. Giese and J.S. Pearse. Academic Press, New York, 5: 113–1295.

Singmorthy, G., 1983. The possible metabolic diversions adapted by the freshwater mussel to counter to toxic metabolic effects of selected pesticides. *Indian J. Comp. Anim. Physiol.*, 1: 95–106.

Swami, K.B., Rao, K.S.J., Reddy, K.S., Chetty, C.S., Indira, K. Shrinivasamoorthy, Vedpathak, A.N., Mulfyand, S.D. and Mane, U.H., 1987. Effect of temperature on respiration biochemical composition in a freshwater bivalve, *Indonaia caeruleusi* (Prashad, 1918). *Proc. Nat. Con. Env. Impact of Biosystem.*

Widdows, J. and Bayne, B.L., 1971.Temperature acclimation of *Mytilus edulis* with reference to its energy budget. *J. Mar. Biol. Assoc.*, U.K., 51: 827–843.

Willingen, C.D., 1969. The effect of *Mytilicola intestinalis* on the biochemical composition of mussels. *J. Mar. Biol. Assoc.*, U.K., 49: 161–173.

Chapter 15

Water Quality Management for Sustainable Aquaculture

☆ *Meenakshi Jindal*

ABSTRACT

Water quality management is vital for sustained high fish production in intensive or semi-intensive aquaculture systems. The artificial feed constitutes the major input component for sustained high production, but also responsible for deterioration of water quality in such systems. The main pollutants in farm effluent water are phosphorous, nitrogen and faecal solids. These can be reduced by using pelleted water stable feed and by supplementation of phytase enzyme in feed along with modifying feeding levels and methods.

Introduction

Global fisheries is facing constant decline in fish stocks, both in coastal and inland water resources on account of continuously increasing water pollution. In intensive culture systems the feed is considered to constitute major input cost and an important source of pollution leading to eutrophication. Faecal solids and excessive presence of phosphorus and nitrogen originating from unconsumed feed and undigested nutrient constituents in faeces are major causes of water pollution in culture systems. Undigested protein and unutilized phosphorus are excreted in feces. The unconsumed feed left at bottom and indigested nutrients in faecal solids release their phosphorus and nitrogen contents after decomposition and increase biological oxygen demand. Reduction in wastage from unconsumed feed is possible by making pellets of feed and increasing water stability of pellets and devising appropriate feeding strategies and methods.

The strategies for reducing amount of unconsumed feed, excretion level of phosphorus and nitrogen, undigested part of faeces, fecal solid discharges *etc.* for controlling pollutants originating from feed are discussed separately.

Faecal Solids

Various undigested organic constituents like protein, lipids, soluble and fibrous carbohydrates *etc.* form solid faecal wastes. These organic solid wastes can be reduced by proper management strategies:

1. Developing diets using ingredients with high nutrient digestibility.
2. Supplementation of diets with microbial enzymes such as amylases and proteases to improve the digestion of carbohydrate and protein resulting in increased nutrient utilization.
3. Supplementation of diets with premix minerals and amino acids. Each Kg of which contains Copper–312mg; Cobalt–45mg; Magnesium–2.114g; Iron–979mg; Zinc–2.13g; Iodine 156mg; DL-Methionine–1.92g; L-lysine mono hydrochloride–4.4g; Calcium 30 per cent and Phosphorous–8.25 per cent (Jindal *et al.*, 2007a, b)

Phosphorus

Phosphorus is an essential mineral which is available only by diet and its dietary requirement for various fish species ranges between 0.3 to 0.9 per cent of diet. Higher dietary phosphorus requirement has been reported for scaly fishes than scale-less one. Fish excrete phosphorus in soluble and particulate forms. The soluble fraction called ortho-phosphate (o-PO_4) is most available for plant growth. However, the main loading of phosphorus to the environment was reported to be via faecal pellets (Figure 15.1).

Various Management Steps in Reducing the Phosphorus Excretion by Fish

1. First step in reducing the phosphorus excretion by fish is to assess the accurate level of bioavailability of phosphorus in feed ingredients. It assumes significance because easiest way in reducing phosphorus excretion is to reduce the proportion lost in faeces. In general feed made from animal by products contain comparatively high phosphorus level coming from bone component. The feed ingredients produced from plants *i.e.* grains and oilseeds, are relatively low in total phosphorus, but it is stored in seeds as phytate phosphorus which is indigestible in monogastric animals like fishes. However, indigestible phytate phosphorus becomes readily available to fish when phytase enzyme is supplemented in diet.
2. Formulation of feed with phosphorus levels at levels required by a fish species at each stage of production. Using most comprehensive approach, 0.66 per cent and 0.55 per cent phosphorus requirements for 200g and 400g rainbow trout respectively were reported.
3. Formulation of feeds to match the available phosphorus level in the feed to the requirement of the fish. When fish are fed with phosphorus deficient diet then phosphorus from body reserves are drawn up and if dietary phosphorus exceeds requirement levels these are absorbed in excess and then excreted in urine (Jindal, 2007).

Nitrogen

In freshwater systems, nitrogen is sometimes a limiting nutrient, so adding it stimulates plant and algal growth. A majority of the excess nitrogen in either tank or pond culture systems originates as ammonia excreted by fish through gills (Kaushik and Cowey, 1991). The ammonia as a waste product, is formed during the breakdown of proteins and excess amino-acids not incorporated into the tissues by the fish (Kibria *et al.*, 1998). Nitrogenous wastes are dietary in origin and depending upon species up to 52–95 per cent of feed nitrogen is excreted as waste (Figure 15.1). Nitrogen excretion is influenced by the quantity and quality of dietary protein. Protein quality is assessed by amino acid

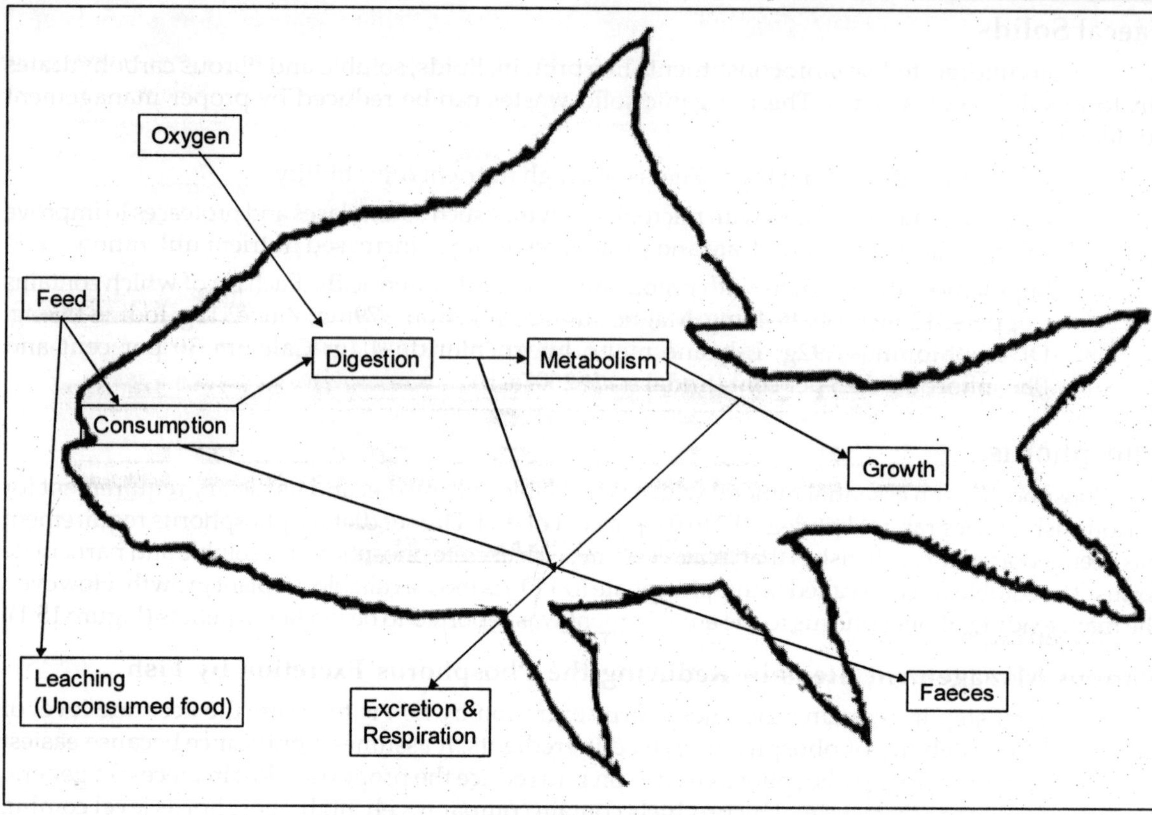

Growth = In–Out

***Feed * Feed loss (Unconsumed feed particles)**

***Oxygen * Faeces**

*** Excretion and Respiration**

Figure 15.1: Pathway Showing Utilization of Dietary Nutrition

composition and digestibility. Dietary protein not digested and absorbed in gastrointestinal tract is excreted in faeces. When the concentration of dietary protein is excessive relative to non-protein energy then the portion of dietary protein not used for protein accretion is broken down and used for energy resulting in elevated ammonia production by fish. Fishes are considered metabolically efficient at utilizing protein for energy and higher level of digestible protein relative to digestible energy increased ammonia excretion in rainbow trout (Figure 15.2).

Various Strategies Properly Managed to Lower Nitrogen Excretion

1. One strategy to lower nitrogen excretion involves increasing digestible energy level in the diet relative to digestible protein. Increasing energy density of diets by lipid supplementation to enhance growth and protein retention has been observed in Atlantic salmon (Kaushik and Cowey, 1991).
2. Most appropriate approach for increasing nutrient density of feed has been suggested to exclude ingredients with low protein and energy contents.

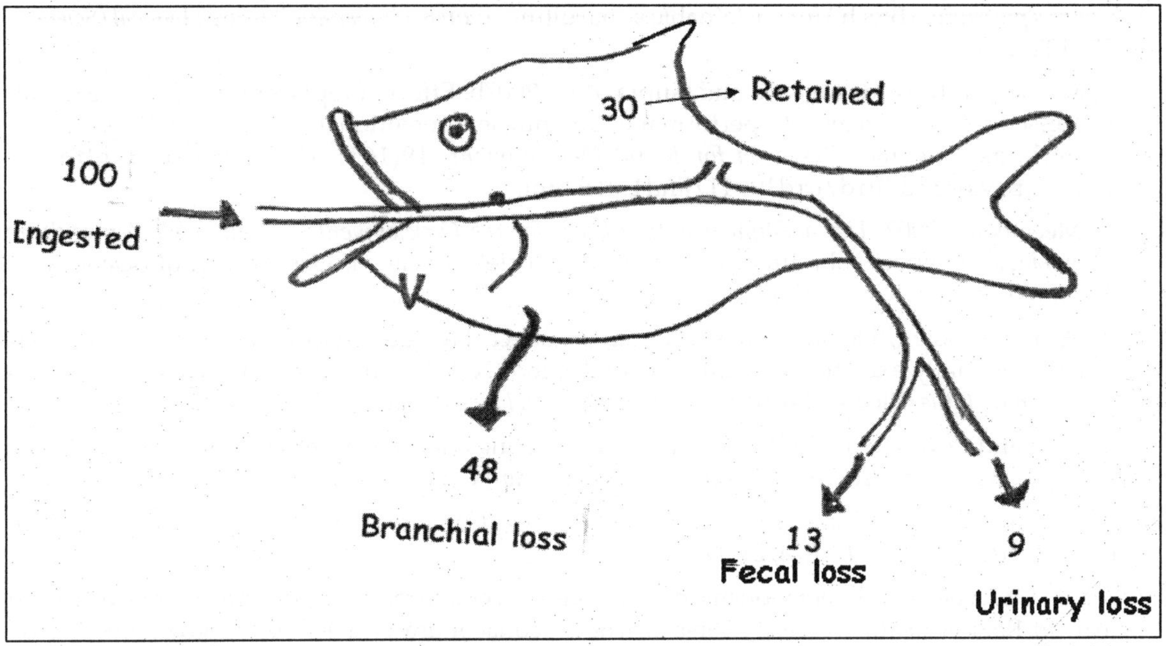

Figure 15.2: Generalized View of Nitrogen Balance in Fish (After Luquet, 1982)

3. Replacing feed ingredients having low protein digestibility with those having high protein digestibility coefficients is another strategy for lowering nitrogen excretion.
4. Manufacturing procedures like use of extrusion processing of fish diets which has been found to increase digestibility of diets and reduce ammonia excretion is another useful tool for curtailing nitrogen excretion (Jindal, 2007).
5. Reducing nitrogen level will require formulation of feed which is able to support high percentage of nitrogen retention (protein increase) by fishes (Jindal *et al.*, 2007a b, Jindal *et al.*, 2008).

Conclusion

Thus it is evident that by using appropriate strategies involving diet composition, feed manufacturing processes, feeding strategies *etc.* level of enriching nutrients in fish rearing systems can be significantly reduced without compromising production or profitability.

Acknowledgement

The author acknowledges funding received under the scheme "Women Scientist Scholarship Scheme for Societal Programmes (WOS-B), Department of Science and Technology, Government of India" for the publication of this research article.

References

Jindal, M., Garg, S. K. and Yadava, N. K., 2007a. Effect of replacement of fishmeal with dietary protein sources of plant origin on the growth performance and nutrient retention in the fingerlings of

Channa punctatus (Bloch.) for sustainable aquaculture. *Punjab University Research Journal (Science)*, 57: 132–138.

Jindal, M., Garg, S.K., Yadava, N.K. and Gupta, R.K., 2007b. Effect of replacement of fishmeal with processed soybean on growth performance and nutrient retention in *Channa punctatus* (Bloch.) fingerlings. *Livestock Research for Rural Development*, 19(165). Retrieved from http: //www.cipav.org.co/lrrd/lrrd19/11/jind19165.htm.

Jindal, Meenakshi, 2007. *Use of Supplementary Feeds for the Development of Sustainable Aquaculture Technology*. A report submitted to Science and Society Division, Department of Science and Technology, New Delhi, pp. 47.

Jindal, Meenakshi, Garg, S.K. and Yadava, N.K., 2008. Effect of replacement of fishmeal with processed soybean on daily excretion of ammonical-nitrogen (NH_4-N) and ortho-phosphate (O-PO_4) in *Channa punctatus* (Bloch). *Punjab University Research Journal (Science)*, 58 (In press).

Kaushik, S.J. and Cowey, C.B., 1991. Dietary factors affecting nitrogen excretion by fish. In: *Nutritional Strategies and Aquaculture Waste*, (Eds.) C.B. Cowey and C.Y. Cho, p. 7–19.

Kibria, G., Nugegoda, D., Fairclough, R. and Lam, P., 1998. Can nitrogen pollution from aquaculture be reduced? *NAGA, ICLARM*, 21: 17–25.

Luquet, P., 1982. Aspects du metabolisme des poisons particulierement importants pour la qualite de leau. In: Razionale utilizzo delle risorse idriche in acquacoltura. Atti del Convengo Internazionale Verona, Salone Acquacoltura, 16 Ottobre 1982. Pubblicazione curate dal Centro regionale tutela e sperimentazione Pesca e Acquacolture dell' E.S.A.V., Italy, pp. 30–38.

Chapter 16

Modified Atmosphere Packaging (MAP) in Fisheries

☆ *S.S. Todkari and M.M. Girkar*

Modified Atmosphere Packaging

Modified atmosphere packaging (MAP) is a technique used for prolonging the shelf-life period of fresh or minimally processed foods. The air surrounding the package is changed to another composition to achieve preservation effect. MAP is used with various types of products, where the mixture of gases used. The package depends on the type of product, packaging materials and storage temperature. Meat and fish product require low gas permeability films whereas for non-respiring products high barrier films are used.

What is MAP?

MAP or modified Atmosphere packaging, means replacing air in a pack with different mixture of gases. A combination of carbon dioxide, nitrogen and oxygen is used for replacing air. The proportion of each component gas is fixed when the mixture is introduced, but there is no further control during storage and the composition of the mixture may change slowly.

Advantages of MAP Technology

- ☆ Increased shelf-life.
- ☆ The appearance of pack is attractive and since transparent packaging is used buyer can clearly see the product.
- ☆ MAP products are odourless, easy to lable and convenient to handle. They are leak proof and robust.
- ☆ Better utilization of labour, space and equipments thus reducing production and storage cost.

Disadvantages of MAP Technology

- ☆ Relatively expensive.
- ☆ High capital cost of packaging machinery and thermo formable film.
- ☆ Modified atmosphere packs are bulkier than other forms of pack and are therefore costlier to carry and store.
- ☆ Potential growth of food borne pathogens due to non-maintenance of required storage temperature.

Gases Used in MAP

In MAP, the pack is flushed with a gas or a mixture of gases. Oxygen, nitrogen and carbon dioxide are commonly used gases. Minimum oxygen level is used to pack food under modified atmosphere because oxygen reacts with foodstuff leading to oxidation. Oxygen also combines with fats and oil causing rancidity. Nitrogen is an inert gas and acts as a cushion thereby preventing pack collapse. It replaces oxygen from the pack and delays oxidative rancidity. Carbon dioxide is responsible for bacteriostatic and fungistatic effect. It retards growth of moulds and aerobic bacteria. A ratio of 3 parts of gas mixture to 1 part of fish by volume is recommended for modified atmosphere packs.

Quality of Fish

Fish with highest quality should be used for modified atmosphere packs in order to gain good storage life. White fish equivalent to 1-4 days in ice and free from blemishes and parasites should be used. Herring and mackerel equivalent to 1-3 days iced storage and salmon and trout used for smoking should also be equivalent to 1-3 days in ice.

Packing the Fish

Fish should be handled hygienically and kept chilled from capture till they are packed. Whole fish and fillets should be kept in ice while smoked products should be held in a chill room till processing.

Wet fish products are likely to exude drip and can be laid on a pad of absorbent paper inside the pack. Packs with faulty seals can be detected by preening them with hands. Packs should be labelled according to existing regulations and should be marked with sell–by date.

Storage Life of Packs

Storage life depends upon the fish used, its initial quality and fat content, the nature of the finished product, temperature of storage and in a modified atmosphere, the gas mixture.

Packs should be stored at 0-5° C. The storage life of herring, mackerel, salmon, trout and smoked fish product is not extended in a modified atmosphere at 0°C.

References

Bharti, A. and Sahoo, J., 1999. Modified atmospheric packaging of meat and meat products: aspects of packaging materials, packaging environment and storage temperature. *Indian Food Ind.*, 18: 299.

Brody, A.L., 2003. Reduced oxygen packaging of seafood. *Food Technol.*, 57(10): 124.

Farber, J.M. and Dodds, K.L., 1995. *Principles of Modified Atmospheric Packaging and Sous Vide Packaging*. Technomic, Lancaster, PA.

Ninawe, A.S. and Rathankumar, K., 2008. *Fish Processing Technology and Product Development*. Narendra Publishing House, Delhi.

Sivvertsvik, M., Jeksrud, W.K. and Rosnesa, J.T., 2002. A review of modified atmospheric packaging of fish and fishery products: Significance of microbial growth, activities and safety. *Inst. J. Food Sci. Technol.*, 37: 107.

Venugopal, V., 2006. *Seafood Processing*. CRC Press.

Chapter 17

Piscivorous Birds of Dhanora Tank in Bhokar Tahasil of Nanded District, Maharashtra

☆ *V.S. Kanwate and V.S. Jadhav*

Introduction

Birds occupy an important position in the animal kingdom, specially in relation to man. Economically, they are both useful and harmful to man's interests *i.e.* as food. In industry, art and ornamentation, as currency, as fertilizer, as pollinators, in biological control, as predators, as scavengers, in medicine, as messengers, as signals, for amusement, Aesthetic value, Injurious birds–menace to agriculture, destroyer of game birds and other animals, pests of fruits and stored grains, spread of disease, pests of honey bees.

By the large, however, the balance sheet on birds has always tipped favourably towards the credit side. Thus the birds are of great economic importance to man, help in controlling population of different pests, scavengers of nature and are also pollinating agents. (Singh, 1929; Ali 1932; Kannan, 1980 and Davidar, 1985). Piscivorous birds have been studied earlier by Ghazi (1962) and Kulkarni (2006).

Dhanora tank is used for irrigation and fish culture activities. Aquatic birds were observed on this tank. The present work was undertaken to do survey and identified of piscivorous birds of this tank. So that, it will helpful to control population of these birds.

Materials and Methods

Dhanora tank is an oldest minor irrigation reservoir constructed near the village Dhanora in 1968 on Bhokar-Nanded road. The water spread area of the tank is 85 hectors. Periphery boundaries of reservoir are having, some Cyprus and other marginal weeds which also provide suitable sites for nesting of these birds. For survey and identification of birds two visits (Season) were done (Date 10.11. 2008, 10.12. 2008, 10.02. 2009 and 10.03. 2009) in morning hours *i.e.* 7.00 to 10.00 am and evening hours 4 to 5.20 pm. Birds were identified at the spot as per the guidelines given by Ali and Ripley (1996) by using binoculars 7x and 8x magnifications.

Results and Discussion

Observed birds are listed in the Table 17.1 on the basis of their common names, scientific names, total counts, nature of abundance and migratory behaviour.

In the present study total 10 species of Piscivorous birds were identified. Out of them 05 are residents, 01 migratory and 04 residential migratory. The species feed on fishes therefore affecting the reservoir fishery. They are also carries of pathogens (Lagler, 1978; Jhingran, 1988) and therefore, it is necessary to reduce their population. This can be done by eradicating aquatic weeds and clearing the peripheral margin of the reservoir.

Table 17.1: Occurrence of Piscivorous Birds of Dhanora Tank, Bhokar, Dist. Nanded (M.S.)

Sl.No.	*Common Name*	*Scientific Name*	*Count*	*Abundance*	*Migratory Behaviour*
1.	Gray Heron	*Ardea cinerea*	10	r	RM
2.	Little Egret	*Egretta garzetta*	05	r	RM
3.	Red Wattied Lapwing	*Vanellus indicus*	05	r	R
4.	Small Blue Kingfisher	*Alcedo athis*	02	r	R
5.	White Breasted Kingfisher	*Halcyon smyrnensis*	03	r	R
6.	Lesser Pied Kingfisher	*Ceryle rudis*	02	r	R
7.	Purple Moor-hen	*Prophyrol porphysio*	02	r	R
8.	Common Coot	*Fulica atra*	102	c	RM
9.	Spot-billed Duck	*Anas poeciiorthyncha*	50	o	RM
10.	Red-creasted Pochard	*Rhodonessa rufina*	12	r	M

Abbreviations used in the list are as follows:

1) For movement: R: Residential; M: Migrant; RM: Residential migrant

2) For Abundance: C: Common (above 100); O: Occasional (above 50); U: (above 20); r: (above and 5).

Acknowledgement

Authors are thankful to Dr. A.N. Kulkarni, Dept. of Zoology, Science College, Nanded for encouragement.

References

Ali, S., 1932. Flower-birds and birds flowers in India. *J. Bom. Nat. Hist. Soc.*, 35: 573–605.

Ali, S. and Ripley, S.D., 1996. *A Pictorial Guide to the Birds of Indian Sub-continent*. BNHS Bombay, pp. 1–172.

Davidar, P., 1985. Ecological interactions between the mistletoes and their avian pollinators in South India. *J. Bom. Nat. Hist. Soc.*, 82: 45–60.

Ghazi, H.K., 1962. Piscivorous birds of Madras. *Madras J. Fisheries*, 1(1): 106–107.

Jhingran, V.G., 1988. *Fish and Fisheries of India*. Hindustan Publishing Co-operation, New Delhi, pp. 1–664.

Kannan, P., 1980. Nectar feeding adaptation of flower birds. *J. Bom. Nat. Hist. Soc.*, 75 (Suppl): 1036–1050.

Kulkarni, A.N. and Kanwate,V.S., 2006. Piscivorous birds of Dongarheda irrigation tank, Dist. Hingoli (M.S.). *J. Aqua. Biol.*, 21(1): 86–87.

Lagler, K.F., 1978. *Freshwater Fishery Biology*, WmC Brownn Comp. Publ., Dubuque, Lowa.

Singh, T.C.N., 1929. A note of the pollination of *Erythring indica* by birds. *J. Bom. Nat. Hist. Soc.*, 33: 460–62.

Chapter 18

Chloride Content in Water from Shikara Dam Near Mukhed in Nanded District

☆ *M.S. Pentewar, V.S. Kanwate and V.R. Madlapure*

Introduction

Chlorides occurs naturally in all types of waters. In natural freshwaters, however, its concentration remains quite 10 w. The most important source to chloride in natural waters is the discharge of sewage. In very high concentration it gives a salty taste to the water. Man and other animal excrete very high quantities of chlorides together with nitrogenous compounds. About 8.15 grams of NaCl is exorted by a person per day. Therefore, the chlorides concentration senses and indicator of pollution by sewage. Industries are also important sources of chloride. Chlorides are highly soluble with most of the naturally occurring action and do not precipitate sedimented and cannot be removed biologically in treatment of the wastes. It is harmless up to 1500 mg/lit. Concentration but produce a salty taste ate 250-500 mg/lit. level.

Materials and Methods

Chloride is estimated on the spot immediately water sample collected from each spot from a depth for 5-feet from surface of water. Chloride content is estimated by using titrametric method described in standard method of water and wastewater APHA, AWWA, Wept (1985) and Trivedi and Goel (1985).

Result and Discussion

From the present investigation study chloride was found to be 17.1 mg/lit. to 38.67 mg/lit. at spot–A and 19.14 to 40.55 mg/lit at spot–B in year 2003-2004 June to August. Maximum chloride concentration was recorded in summer while minimum in monsoon.

Minimum range in monsoon may be due to dilution of monsoon floods and maximum in summer due to increase in evapotranspiration. The same results reported by Ansari (1993).

The highest values 38.67 was recorded in the month do may at spot–A and 40.55 in April at spot–B. While lowest value 17.1 in September at spot–A and 19.14 in August at spot–B values of spot–B is higher than spot–A, may be due to man made pollution near the dam, showing below in Table 18.1.

Table 18.1: Table Showing Monthly Values of Chlorides in mg/lit

Monthly	*Year*	*Spot–A*	*Spot–B*
June	2003	22.9	23.56
July	2003	18.5	20.48
August	2003	17.66	19.14
September	2003	17.1	19.59
October	2003	18.33	20.37
November	2003	20.5	22.86
December	2003	23.38	25.74
January	2004	27.5	28.13
February	2004	37.28	39.98
March	2004	38.24	39.67
April	2004	38.52	40.55
May	2004	38.67	39.12
June	2004	22.09	23.56
July	2004	18.5	20.48
August	2004	17.66	19.14

Pandarkar (1998), found the annual range of chloride to be 709 to 84.77 mg/lit. minimum value was in winter and maximum during mansoon season. Mane (2002), noted chloride range from 39 to 84.1 mg/lit. maximum in summer. While minimum in mansoon. The recorded values of chloride in present study was similar to the above scintic the water chloride range of Sikara dam is suitable for the fish life is confirmed.

References

Ansari, M.A., 1993. Hydrobiology studies of Godawari River water at Nanded. *Ph.D. Thesis*, Marathwada University Aurangabad, M.S.

APHA, AWWA, WPCF, 1985. *Standard Method for the Examination of Water and Wastewater*, 16th Edn., p. 1–12, 34.

Mane, A.M., 2002. Study of hydrobiology of Manar river, Degloor, Dist. Nanded. *Ph.D. Thesis*, S.R.T.M.U., Nanded, M.S.

Pandarkar, A.K., 1998. Some aspects of the biology of freshwater fish *Macrones bleekeri* from a lake near Ahmadnagar. *Ph.D. Thesis*, Dr. B.A.M.U., Aurangabad, M.S.

Trivedi, R.K. and Goel, P.K., 1995. *Chemical Biological Methods for Water Pollution Studies*. Environmental Publication, Karad, pp. 16.

Chapter 19

Studies on Aquatic Insects in Relation to Physico-chemical Parameters of Anjani Reservoir from Tasgaon Tahsil of Sangli District, Maharashtra

☆ *S.A. Khabade and M.B. Mule*

Introduction

Water quality plays an important role in the growth of aquatic animals and their distribution and abundance. The water quality standards below and above the optimum level may lead to either stress or death among the aquatic animals. The water quality mainly depends on physical, chemical and microbiological parameters. Sometimes, water quality may be changed due to organic and chemical pollution of water body.

Aquatic insects comprise an ecologically important group of organisms in freshwater systems. They are known to play a very important role in the processing and cycling of nutrients, as they belong to several specialist feeding groups such as filter feeders, deposit collectors, scrapers, shredders as well as predators (Lamberti and Moore, 1984). Their importance as biomonitor or indicators of freshwater pollution has also been amply demonstrated (Wiederholm, 1984; Metcalfe, 1989).

Aquatic insects spend part of all their lives in soil or water exhibit special structural and behavioural adaptations to the physico-chemical and biotic conditions found in each. These insects form an important link in the nutritional cycle of an aquatic ecosystem, as they constitute a bulk of food for fishes. Extensive work has been done on the seasonal hydrobiological conditions of freshwater bodies of India (Sinha and Sinha, 1999; Kaushik *et al.*, 1990; Pandey, *et al.*, 1992; Singh, 1993; Arvind Kumar, 1994).

Most of the waterbodies have been unmindfully used for the disposal of waters as dust-bin far beyond their assimilative capacities and have been grossly polluted. The domestic sewage and industrial effluents contain pollutants that cause harmful effect on receiving waters and adverse impact on human health as well as aquatic biota (Telliard and Robin 1987; Kumar 1996b).

Since aquatic insects are bound to an aquatic habitat for most part of their life-cycles, any change in their number and composition in the population at a given time and space may indicate a change in the water quality. Benthic insects are considered as promising organism for use in diversity biomonitoring because of their case of collection, large number of species and sensitivity to water quality (Roy and Sharma, 1983; Kumar 1995a). As such many of aquatic insects form biological indicators of the environmental quality. Thus, by studying them it is possible to anticipate the impact of pollution even before drastic physical and chemical changes have occurred.

Information on the inter-relationship between aquatic insect population and physico-chemical conditions is scanty. Hence, the influence of some physico-chemical factors on the abundance of aquatic insects of Anjani reservoir of Tasgaon tahsil of Sangli district has been studied.

The environmental pollution affects the general quality of our surroundings and poses risk to our health and wellbeing. The contaminants present in the earth in one way or other get collected through the streams and rivers and ultimately reach the ocean. Making assessment of water quality and devising control strategies are not static, but an ongoing process. A number of factors act concurrently in a lake and in turn on the quality of the water (Churchill and Buckingham 1956). Deshmukh (1964) has studied the physico-chemical characteristics of Ambazari lake in Nagpur, Maharashtra. The water quality and conservation aspects of five water bodies in and around Hyderabad, Andhra Pradesh, are discussed by Kodarkar (1995). Similarly, a number of studies on physico-chemical and biological quality of the waters have been extensively carried out (Busulu *et al.*, 1967; Chakraborty *et al.*, 1977; Adwant, 1989; Khatavkar and Trivedy, 1992; Joshi and Bisht, 1993 and Gill *et al.*, 1993). Considerable work has been carried out on water quality assessment of freshwater bodies by using the species diversity in other regions of India (Sharma and Rai 1991; Karim 1993).

Materials and Methods

The aquatic insects were collected monthly during July 2004 to June 2005 period from the Anjani water reservoir with an insect collecting net made up of nylon cloth having mesh size: 40-80 cm^2. The samples were cleaned and preserved in 5 per cent formalin. Then the insects were identified in the laboratory with the help of standard literature of Tonapi (1959); Michael (1973), Macan (1959).

Monthly collection of water sample was done from the period July 2004 to June 2005, by using samplers *i.e.* plastic containers of 5 litre size. The sites selected for water sampling are site I (SI) and site II (S II) which lies near the dams earthen embankment and the feeders canal which recharge the reservoirs water respectively. The meterological parameters such as air temperature and humidity were determined in the field. Similarly, few physico-chemical parameters like water temperature, pH and dissolved oxygen were determined in the field during monthly visits. The electrical conductivity, total alkalinity, hardness, magnesium, calcium, chlorides, residual chlorine, acidity, free CO_2, hydrogen sulphide, sodium, potassium, nitrate and phosphate were analysed in laboratory, using standard methods of water analysis described by Trivedy and Goel (1984) and APHA, AWWA, WPCF (1985).

Results and Discussion

The results on water quality assessment in terms of a number of physico-chemical parameters are summarized in the Tables 19.3 and 19.4.

Table 19.1: Monthwise Record of Rainfall in mm During July 2004 to June 2005 of the Savlaj Block Under which Anjani Water Reservoir Comes

Sl.No.	*Month*	*Year*	*Rainfall in mm*	*Name of the Block*	*Water Reservoirs Coming Under Savlaj Block*
1	July	2004	59.6		Siddhewadi
2	Aug.	2004	75.00		Anjani
3	Sept.	2004	94.20		Balgawade
4	Oct.	2004	14.00		Bastawade
5	Nov.	2004	—	Savlaj Block	
6	Dec.	2004	—		
7	Jan.	2005	—		
8	Feb.	2005	—		
9	Mar.	2005	—		
10	April	2005	—		
11	May	2005	—		
12	June	2005	14.00		

—: No Rainfall.

Table 19.2: Solubility of Oxygen in Pure Water Exposed to Water-Saturated Air at Mean Sea Level Pressure of 760mm Hg.

Temperature (°C)	*DO (mg/liter)*	*Temperature (°C)*	*DO (mg/liter)*
0	14.16	18	9.18
1	13.77	19	9.01
2	13.40	20	8.84
3	13.05	21	8.68
4	12.70	22	8.53
5	12.37	23	8.38
6	12.06	24	8.25
7	11.76	25	8.11
8	11.47	26	7.99
9	11.19	27	7.86
10	10.92	28	7.75
11	10.67	29	7.64
12	10.43	30	7.53
13	10.20	31	7.42
14	9.98	32	7.32
15	9.76	33	7.22
16	9.56	34	7.13
17	9.37	35	7.04

Source: Handbook of Common Methods in Limnology by Lind, O.T., 1974.

Table 19.3: Meterological Parameters of Anjani Water Reservoir from July 2004 to June 2005

Sl.No.	Parameters	Month and Year																							
		Jul 2004		Aug 2004		Sept 2004		Oct 2004		Nov 2004		Dec 2004		Jan 2005		Feb. 2005		Mar 2005		April 2005		May 2005		June 2005	
	Sites	S_1	S_2	S_1	S_2	S_1	S_2	S_1	S_2	S_1	S_2	S_1	S_2	S_1	S_2	S_1	S_2	S_1	S_2	S_1	S_2	S_1	S_2	S_1	S_2
1.	Air Temperature (°C)	31	31	32	33	33	34	31	32	29	28	27	26	26	26	30	30	32	32	37	36	35	38	32	31
2.	Humidity %	72	67	72	65	73	62	52	52	52	45	52	52	50	31	73	79	73	46	36	41	67	65	55	55

All values are mean of four readings.

Table 19.4: Physical Parameters of Anjani Reservoir from July 2004 to June 2005

Sl.No.	Parameters	Jul 2004		Aug 2004		Sept 2004		Oct 2004		Nov 2004		Dec 2004	
	Sites	S_1	S_2	S_1	S_2	S_1	S_2	S_1	S_2	S_1	S_2	S_1	S_2
1.	Water Temperature (°C)	30	30	28	29	27	28	28	28	26	26	25	24
2.	Transparency (cm)	19.5	16	28.5	20.5	67	71	136	122.5	78.8	66.5	95.5	89
3.	Total solids) (mg/l)	3890	2390	3700	1967	1080	1295	910	700	778	890	1300	1100
4.	TDS (mg/l)	680	1160	1268	1600	840	783	560	400	336	400	763	950
5.	TSS (mg/l)	3210	1230	2432	367	240	512	350	300	442	490	537	150

Sl.No.	Parameters	Jan 2005		Feb 2005		Mar 2005		Apr 2005		May 2005		Jun 2005	
	Sites	S_1	S_2	S_1	S_2	S_1	S_2	S_1	S_2	S_1	S_2	S_1	S_2
1.	Water Temperature (°C)	24	24	27	28	28	29	31	31	30	29	29	30
2.	Transparency (cm)	74.5	83	106.5	117.5	94.35	98.3	47.2	47.6	45.5	46.0	22.6	19.0
3.	Total solids (mg/l)	850	940	1095	1135	1888	2091	3350	4830	5215	3393	3514	3920
4.	TDS (mg/l)	690	840	385	407	1233	1115	1093	2021	2814	2112	3200	3140
5.	TSS (mg/l)	160	100	710	728	655	976	2257	2809	2401	1281	314	780

All values are mean of four readings.

Table 19.5: Chemical Parameters of Anjani Reservoir from July 2004 to June 2005

Sl.No.	Parameters	Jul 2004		Aug 2004		Sept 2004		Oct 2004		Nov 2004		Dec 2004	
	Sites	S_1	S_2	S_1	S_2	S_1	S_2	S_1	S_2	S_1	S_2	S_1	S_2
1.	pH	9.2	9.4	9.0	9.0	8.5	8.6	8.7	8.7	8.5	8.5	8.7	8.7
2.	Electrical conductivity (mmhos)	0.359	0.353	0.378	3.360	0.404	0.393	0.468	0.472	0.501	0.413	0.523	0.601
3.	Total alkalinity	240	220	140	130	335	345	330	325	435	395	360	375
4.	Acidity	Ab	Ab	Ab	Ab	Ab	Ab	Ab	Ab	Ab	Ab	Ab	Ab
5.	Hardness	120	130	106	104	158	156	146	146	142	150	124	140
6.	Magnesium	16.55	18.01	10.71	10.71	22.40	20.45	19.96	20.94	18.99	19.96	11.68	13.14
7.	Calcium	20.04	22.44	24.84	24.04	26.45	28.85	25.65	24.4	25.65	27.25	30.46	34.46

Contd...

Table 19.5–Contd...

Sl.No.	Parameters	Month and Year											
		Jul 2004		Aug 2004		Sept 2004		Oct 2004		Nov 2004		Dec 2004	
	Sites	S_1	S_2	S_1	S_2	S_1	S_2	S_1	S_2	S_1	S_2	S_1	S_2
8.	Chloride	25.56	21.3	28.4	29.82	26.98	22.72	15.62	17.04	25.56	24.14	11.36	9.94
9.	Residual Chlorine	Ab	Ab	Ab	Ab	Ab	Ab	Ab	Ab	Ab	Ab	Ab	Ab
10.	Dissolved O_2	4.83	6.04	7.04	7.24	8.85	9.06	8.45	10.06	8.25	9.66	9.46	9.66
11.	Free CO_2	Ab	Ab	Ab	Ab	Ab	Ab	Ab	Ab	Ab	Ab	Ab	Ab
12.	Hydrogen Sulphide	1.84	1.41	0.84	0.56	1.69	1.13	0.28	0.42	1.13	0.84	0.28	1.84
13.	Sodium	23.00	22.00	21.00	20.00	19.80	19.50	20.60	19.20	24.00	22.30	36.50	34.00
14.	Potassium	2.1	2.3	1.9	1.8	2.0	1.9	2.2	2.3	3.1	3.8	5.0	4.9
15.	Nitrate	24.00	23.00	29.00	28.00	31.00	28.00	34.00	32.00	28.00	24.00	28.00	30.00
16.	Phosphate	0.080	0.09	0.11	0.09	0.12	0.14	0.18	0.09	0.14	0.17	0.08	0.09

Table 19.5: Chemical Parameters of Anjani Reservoir from Jan 05 to June 2005

Sl.No.	Parameters	Month and Year											
		Jan 2005		Feb 2005		Mar 2005		Apr 2005		May 2005		Jun 2005	
	Sites	S_1	S_2	S_1	S_2	S_1	S_2	S_1	S_2	S_1	S_2	S_1	S_2
1.	pH	8.7	8.7	8.4	8.4	8.5	8.6	8.6	8.6	8.7	8.6	8.6	8.6
2.	Electrical conductivity (mmhos)	0.341	0.322	0.422	0.523	0.498	0.467	0.403	0.507	0.514	0.500	0.402	0.385
3.	Total alkalinity	330	335	275	250	345	355	380	380	235	240	335	450
4.	Acidity	Ab	Ab	Ab	Ab	Ab	Ab	Ab	Ab	Ab	Ab	Ab	Ab
5.	Hardness	68	70	136	152	148	138	158	168	198	202	150	162
6.	Magnesium	1.46	1.46	14.12	18.99	20.94	19.48	23.37	26.29	30.68	31.16	18.01	19.96
7.	Calcium	24.84	25.65	31.26	29.65	24.84	23.24	24.84	24.04	29.85	29.65	30.46	32.06
8.	Chloride	12.78	14.2	17.04	18.46	21.3	25.56	28.4	32.66	36.92	38.34	19.88	18.46
9	Residual Chlorine	Ab	Ab	Ab	Ab	Ab	Ab	Ab	Ab	Ab	Ab	Ab	Ab
10.	Dissolved O_2	9.26	9.26	8.65	8.65	9.66	10.67	6.24	6.24	6.64	6.84	6.44	6.24
11.	Free CO_2	Ab	Ab	Ab	Ab	Ab	Ab	Ab	Ab	Ab	Ab	Ab	Ab
12.	Hydrogen Sulphide	2.12	2.12	1.69	1.55	0.28	0.70	1.98	2.26	2.26	1.98	0.70	0.99
13.	Sodium	31.00	28.00	33.00	36.00	29.50	31.20	35.50	34.00	39.00	42.20	29.1	30.00
14.	Potassium	4.3	3.8	5.3	5.4	4.4	4.8	5.8	5.2	4.9	3.9	3.1	3.8
15.	Nitrate	29.00	27.00	31.00	34.00	32.00	30.00	32.00	34.00	33.00	35.00	28.00	29.00
16.	Phosphate	0.07	0.13	0.18	0.19	0.12	0.13	0.15	0.16	0.18	0.19	0.13	0.16

Table 19.6: Monthly Occurrence of Aquatic Insects of Anjani Reservoir (from July 2004 to June 2005)

Sl.No.	Insect Orders, Family, Genus and Species	Month and Year											
		July 2004	Aug 2004	Sept 2004	Oct 2004	Nov 2004	Dec 2004	Jan 2005	Feb 2005	March 2005	April 2005	May 2005	Jun 2005
1.	Euphemeroptera												
	Caenidae												
	Caenis sp.	+	+	+++	–	–	–	–	–	–	–	+	+
	Baetidae												
	Cloeon sp.	+	+	+++	++	–	++	++	++	+	–	–	–
	Baetis rhodani	++	+	+++	++	+	++	–	++	–	–	–	–
2.	Hemiptera												
	Notonectidae												
	Anisops sareda	+	++	++	++	+++	++	++	—	+++	+++	++	+++
	Nepidae												
	Ranatra elangdata	+	++	++	++	+++	++	++	+++	+++	+++	++	++
	Laccotrephes maculates	++	++	++	++	++	++	++	+	–	+++	++	+++
	Vellidae												
	Microvelia diluta	++	++	++	++	++	–	–	–	–	–	–	–
3.	Diptera												
	Chironomidae												
	Chironomus sp.	++	++	++	++	–	++	–	–	–	–	–	+
	Culicidae												
	Mosquito sp.	++	++	++	++	++	–	++	++	++	++	++	++
4.	Coleoptera												
	Hydrophilidae												
	Cybister limbatus	++	++	++	++	+	+++	++	+++	+	–	–	–
	Hydrophilus olivaceous	++	++	++	+	+++	+++	++	+++	+	–	–	–
	Regimbartia attenuta	++	++	++	++	+++	+	+++	+++	–	+	+	++
5.	Orthoptera												
	Scellemina harpago	++	++	++	++	–	–	–	–	–	–	–	–

+: Less abundant; ++: Abundant; +++: More abundant; –: Absent.

In the present investigation it has been found that the species diversity was higher during monsoon when rainfall was maximum. Table 19.1 shows the rainfall data during July 2004 to June 2005. During rainy season water temperature was also optimum which affect the species diversity and it becomes higher. During summer, in the month April 2005 when water temperature was high about 31°C, the insect species diversity was minimum in this month. Rao (1976) also opined that a heavy rainfall in monsoon period increases the diversity while the poor rainfall has adverse effect on the diversity index of hemipteran insects. Julka (1977) has also suggested that in the complexes of interdependent factors governing the seasonal variations in diversity of aquatic bugs, temperature and rainfall appear to be the important factors. In the present investigation it is also observed that the species diversity was minimum during summer. In the situation, only pollution tolerant aquatic insects will be present and pollution intolerant species decline. The pollution tolerant species can grow more rapidly without competition for space, nutrients, predation and other extrinsic and intrinsic factors too. This results in heavy dominance of these species leading to the decline in the values of species diversity and also in the evenness of species (Cairns, 1977).

In present study, the rainfall was recorded during July 2004 to October 2004 and also during June 2005. The higher rainfall about 94.2 mm was reported during September 2004. In the remaining months no rainfall was recorded in the Savlaj Block. The air temperature and humidity reported during rainy season was optimum. The maximum air temperature of about 38°C was reported during month May 2005 and minimum air temperature of about 26°C was reported during months December 2004 and January 2005. The maximum humidity 79.00 per cent was reported during February 2005 while minimum humidity 31.00 per cent was reported during January 2005. The optimum range of air temperature and humidity was also responsible for maximum insect species diversity during rainy season.

Light transparency of water depends on the total solids, total dissolved solids and total suspended solids of the water. In present study it is found that these physical parameters not cause any adverse effect on the distribution and abundance of insect diversity, in Anjani water reservoir during study.

In present investigation the maximum water temperature about 31°C was recorded during April 2005 and minimum water temperature about 24°C was recorded during December 2004 and January 2005. The maximum light transparency about 136.0 cm was recorded during October 2004 while minimum light transparency about 16.0 cm was recorded during July 2004. The maximum total solids reported was 5215 mg/L, during May 2005 while minimum total solids reported was 700 mg/L, during October 2004. The maximum total dissolved solids reported was 3200 mg/L during June 2005 while minimum total dissolved solids reported was 336 mg/L during November 2004. The maximum total suspended solids reported was 3210 mg/L during July 2004 while minimum total suspended solids reported was 100 mg/L during January 2005. The analysis of historical records reveals a number of inter-relationship among key water quality parameters used in the assessment of cumulative impacts. The recommended concentration of TSS is 80 mg/L.

In present study it is also found that the chemical parameters such as pH of water, electrical conductivity, total alkalinity, hardness, magnesium, calcium, chloride, dissolved oxygen, acidity, residual chlorine, free CO_2, hydrogen sulphide, sodium, potassium, nitrate, phosphate may not cause any adverse effect on the distribution and abundance of insect fauna.

References

Adwant, M.P., 1989. Limnological studies on Godavari basin at Nanded, Maharashtra, India. *Ph.D. Thesis*, Marathwada University, Aurangabad.

APHA, AWWA, WPCF, 1985. *Standard Method for the Examination of Water and Wastewater*, 16th edn. American Public Health Association Inc., New York, p. 1268.

Busulu, K.R., Arora, H.C. and Aboo, K.M., 1967. Certain observations on self purification of Kher river and its effect on Krishna river. *Ind. J. Environ. Hlth.*, 9(4): 275–296.

Cairns, J.J., 1977. Indicator species by the concept of community structure as amindes of pollution. *Water Resources Bulletin*, 10: 338–347.

Chakraborty, R.D., Roy, P. and Singh, S.S., 1977. A quantitative study on plankton and physico-chemical conditions of river Jamuna at Allahabad in 1945. *Ind. J. Fish.*, 6(1): 186–203.

Churchill, M.A. and Buckingham, R.A., 1956. Statistical method for analysis of stream purification capacity. *Sewage Ind. Wastes*, 28: 517–537.

Deshmukh, S.B., 1964. Physico-chemical characteristics of Ambazari lake water, Nagpur, Maharashtra. *Ind. J. Environ. Hlth.*, 6(3): 166–188.

Gill, S.K., Sahota, G.P.S. and Sahota, H.S., 1993. Phytoplankton and physico-chemical parameters, examination of river Sutlej. *Ind. J. Environ. Pro.*, 13(3): 171–175.

Joshi, B.D. and Bisht, R.E., 1993. Some aspects of physico-chemical characteristics of western Ganga Canal near Jwalapur, Haridwar. *Himalayan J. Env. Zool.*, 7(1): 76–82.

Julka, J.M., 1977. On possible fluctuation in the population of aquatic bugs in a fish pond. *Oriental Insects*, 11: 139–149.

Karim, S.W., 1993. Use of species diversity in the evaluation of water quality of two ponds. *Columban J. Life Sci.*, 1: 27–29.

Kaushik, S., Sharma, S.M.N. and Saksena, D.N., 1990. Abundance of insects in relation to physico-chemical characteristics of pond water of Gwalior (M.P.). *Proc. National Acad. Sci., India*, 60B: II.

Khatavkar, S.D. and Trivedy, R.K., 1992. Water quality parameters of river Panch Ganga near Kolhapur and Ichalkaranji, Maharashtra, India. *J. Toxicol. Env. Monit.*, 2(2): 113–118.

Kodarkar, M.S., 1995. *Conservation of Lakes*. IAAB Publ., Hyderabad, 2: 82.

Kumar, Arvind, 1994. Role of species diversity of aquatic insects in the assessment of population in wetlands of Santhal Parganas (Bihar). *J. Environment and Pollution*, 1(3 and 4): 117–120. *J. of Freshwater Biol.*, 8: 241–146.

Kumar, A., 1995a. Population dynamics and species diversity of odonate larvae in the wetlands of Santhal Pargana, India. *Proc. Nat. Sci.*, 65: 265–278.

Kumar, A., 1996 b. Impact of industrial pollution on the population status of *Plantanista gangetica* in the river Ganga in Bihar, India. *Pol. Arch. Hydrobiol.*, 43: 469–476.

Lamberti, G.A. and Moore, J.W., 1984. Aquatic insects as primary consumers. In: *The Ecology of Aquatic Insects*, (Eds.) V.H. Resh and Rosenberg, D. M. Praeger Publishers, New York, p. 164–195.

Macan, T.T., 1959. *A Guide to Freshwater Invertebrate Animals*. Longman Group Ltd., London.

Metcalfe, J.L., 1989. Biological water quality assessment of running waters based on macroinvertebrate communities: History and present status in Europe. *Env. Poll.*, 60: 101–139.

Michael, R.G., 1973. A guide to the freshwater organisms. *J. Madurai Univ. Suppl.*, 7: 23–36.

Pandey, B.N., Jha, A.K. and Lal, R.N., 1992. Benthic macro-invertebrate communities as indicators of pollution in river Mahanand Katihar. *Oikossy*, 9(1 and 2): 35–29.

Rao, T.K.R., 1976. Bioecological studies on some aquatic Hemiptera. Nepidae. *Entomon.*, 1: 123–132.

Roy, S.P. and Sharma, U.P., 1983. Studies on the role of insects in freshwater ecosystem. In: *Proc. Symp. in. Ecol. and Environ. Manages*, pp. 187–191.

Sharma, U.P. and Rai, D.N., 1991. Seasonal variation and species diversity of Coleopteran insects in fish pond of Bhagalpur.

Singh, U.N., 1993. Studies on food and feeding behaviour of selected aquatic insects in artificial habitat. *J. Comp. Physiol. Ecol.*, 18(2): 69–71.

Sinha, K.K. and Sinha, D.K., 1999. Seasonal variations in biomass and production of aquatic insects in Derlict pond and a managed fish pond of Munger, Bihar. *Env. and Ecol.*, 8(4): 1231–1234.

Telliard, W.A. and Rubin, M.B., 1987. Control on pollutants in wastewater. *J. Chromatographic Science*, 25: 322–327.

Tonapi, G.T., 1959. Studies on the aquatic insects fauna of Poona (Aquatic–Heteroptera). In: *Proc. Nat. Inst. Sci., India*, 25(6): 321–332.

Trivedy, R.K., Goel, P.K. and Trisal, C.L., 1987. *Practical Methods in Ecology and Environmental Science*. Enviro Media Publication, Karad.

Weiderholm, T., 1984. Responses of aquatic insects to environmental pollution. In: *The Ecology of Aquatic Insects*, (Eds.) V.H. Resh, V.H. and D.M. Rosenbery. Praeger Publishers, New York, p. 508–557.

Chapter 20

Physico-chemical Analysis of Small Reservoir Budha Talab in Raipur

☆ *P. Biswas, H.K. Vardia and A. Ghosh*

ABSTRACT

Budha Talab is the largest pond (30.25 ha.) among total 22 ponds present in the Raipur city. It receives domestic waste and sewage and also used for cloth washing, animal bathing and therefore water quality is fast deteriorating. The physico-chemical and of water in Budha Talab were done in the year 2005-2006 from July to June. Out of 13 physico-chemical characters, pH, transparency, dissolved oxygen, alkalinity, hardness, BOD, COD, nitrite nitrogen, nitrate nitrogen, ammonical nitrogen, total orthophosphate and varied significantly throughout the year except dissolved organic matter. The data were analyzed using simple CRD. High BOD and COD indicate increasing organic pollution in the pond. Nitrogen, phosphorus and dissolved organic matter were also found to be high indicating eutrophic nature of the water body.

Introduction

Reservoirs which are situated near human settlements are greatly influenced with the habitation *viz.* with disposal of sewage, soap and detergents and industrial wastes. In developing countries like India where only few of the big cities posses sewage treatment plants, the proper disposal of sewage is an acute problem. Most of the city sewage and domestic water is disposed into rivers, reservoirs and ponds without any treatment.The sewage water is rich in phosphate due to extensive use of hard and/or soft detergents. It is estimated that about 800×10^6 gallons/day of sewage and sullage can yield an annual out turn of 60,000 t NPK. Domestic sewage contains 250-400 ppm Organic Carbon and 80-120 ppm total nitrogen with a C: N ratio of 3: 1 (Mukhophodhyay and Sarangi, 2006). Even small addition of organic matter eutrophicates reservoirs and stimulates bloom in the chain of waters connected to them (Sreenivasan, 1969). Raipur is known as city of ponds. Once upon a time there were

70 ponds (1977) in Raipur city (Marothia, 1997) while reduced to 40 in 1997, but as of today there exists only 22 ponds. Vivekanada sarovar (popularly known as Budha Talab) is the largest tank in the city of Raipur (21°14′N, 81°38′E). It is surrounded by thick dense habitation. Therefore, Physico-chemical and biological conditions of water get affected due to human interference.

Materials and Methods

Surface water samples were collected at monthly intervals for twelve months (July 2005-June 2006) from three sampling sites between 2 p.m. to 5 p.m. from Budha Talab. Water samples were brought to the laboratory and used for the analysis of water temperature, pH, transparency, dissolved oxygen (DO), alkalinity, total hardness, biochemical oxygen demand (BOD), chemical oxygen demand (COD), ammonical-N, Nitrate-N, Nitrite-N and ortho-phosphate. Were carried out according to APHA (1998).

Results and Discussion

Temperature is one of the most important factors in an aquatic environment (Ruttner, 1963). It influences the physico-chemical and biological characters of water body.

The mean water temperature varied significantly through out the year. The highest temperature was recorded in the month of May (35.5 °C) and the lowest was recorded in the month of December (22.3°C). Raipur experiences mild winter and extremely hot summer, which is also seen in the meteorological data collected at Raipur by the Department of Agrometeorology, IGAU, Raipur. Weekly and monthly variations in air temperature show that monsoon starts by mid of June and prolongs from July to September. Winter from October to December, spring from January to March and summer from April to June. Therefore, the seasonal variations in water temperature were also found significant and such variations are bound to influence other water characters, which are depicted in the following part of study. A significant correlation was found between water temperature and total orthophosphate. The pH of Budha Talab water ranged between 7.9 to 9.06 indicating alkaline nature of water throughout the year. Many workers reported pH to normally vary between 7-10 in inland water bodies (Zutshi and Vass 1984; Chari, 1985; Khatavkar and Trivedy, 1993). The highest pH was recorded in the month of October and the lowest in the month of June.

Absence of free CO_2 in Budha Talab may be attributed to the higher pH (above 8.0) as reported by Hutchinson (1957) and Khatavkar and Trivedy (1993). Golterman (1970) while analyzing natural waters for relation between pH and percent of total CO_2 as free CO_2, HCO_3 and CO_3 found that an increased pH means higher carbonate values. Contrary to this, the present study found higher pH values and decreased free CO_2 values. pH was also found to have negative correlation with transparency indicating better growth of plankton. Domestic effluents and surface run off also increase pH as reported by Mohanty (1999).

The mean secchi disc reading varied significantly over the year and ranged between 23-38.1 cm. The highest transparency was in December and the lowest in October. The secchi disc transparency depends upon the suspended particles present in water and therefore, it is a good index of phytoplankton population, especially in productive waters. (Khatavkar and Trivedy, 1993). Selot (1977) observed direct relationship of transparency with primary productivity. A positive significant correlation also found with dissolved oxygen, hardness and nitrate nitrogen.

Dissolved oxygen (DO) in the water depends upon several physico-chemical and biological factors. DO in the surface water of Budha Talab ranged between 5.2-10.3 mg/l (Table 20.1). DO was

Table 20.1: Variation in Mean Physico-chemical Parameters of Water in Budha Talab, Raipur (2005-06)

Parameter	July	Aug	Sep	Oct	Nov	Dec	Jan	Feb	Mar	April	May	June	Mean	SEm	CV (%)
Water temperature (°C)	28.75	33.75	29.75	27.50	23.00	22.50	23.75	28.50	31.25	33.50	35.50	29.50	28.94	±0.20	1.22
Transparency (cm)	36.83	33.83	26.17	23.00	34.67	38.17	37.33	35.33	33.67	30.33	35.33	34.67	33.28	±1.32	6.92
Total hardness (mg/l)	136.50	105.50	120.00	130.00	115.50	135.25	124.00	139.00	132.50	180.50	148.50	178.50	137.15	±6.21	7.84
Alkalinity (mg/l)	147.50	84.00	96.00	99.50	104.00	85.50	96.00	96.00	100.00	113.50	115.50	144.00	106.79	±3.62	5.98
Dissolved oxygen (mg/l)	9.20	5.20	8.90	8.10	7.80	10.30	9.05	9.35	8.10	9.10	9.50	7.85	8.54	±0.20	4.24
pH	8.47	8.45	8.87	9.07	8.37	8.87	8.13	8.97	7.97	8.27	8.73	7.90	8.51	±0.10	2.12
BOD (mg/l)	38.50	58.00	50.00	78.50	78.50	68.00	79.50	98.00	177.50	111.50	74.00	65.00	81.42	±3.18	6.90
COD (mg/l)	71.00	82.00	93.50	137.00	143.00	197.50	171.00	266.50	235.00	155.50	117.00	129.00	149.83	±5.69	6.66
Dissolved organic matter (mg/l)	2.15	2.20	2.50	2.50	2.20	2.35	2.30	2.10	2.05	2.05	1.95	2.25	2.22	±0.15	12.06
Nitrite nitrogen (mg/l)	0.03	0.02	0.08	0.08	0.07	0.03	0.11	0.17	0.09	0.08	0.05	0.03	0.07	±0.02	37.35
Nitrate nitrogen (mg/l)	0.61	0.27	0.40	0.32	0.44	0.26	0.60	0.86	0.40	0.42	0.24	0.34	0.43	±0.08	38.62
Ammonical nitrogen (mg/l)	1.90	2.45	4.80	2.70	3.60	4.30	4.85	3.50	3.60	2.70	2.95	1.55	3.24	±0.31	17.63
Total ortho-phosphate (mg/l)	0.44	0.55	0.85	0.26	0.29	0.23	0.44	0.92	0.88	1.28	0.71	0.38	0.60	±0.01	5.41

Table 20.2: Coefficient of Correlation Between Different Physico-chemical and Biological Parameters (Monthly)

Parameters	Water Temperature	Turbidity	Total Hardness	Total Alkalinity	Dissolve Oxygen	pH	BOD	COD	DOM
Water temperature (°C)	1	–0.1408	0.304704	0.203178	–0.21703	0.070068	0.171218	–0.28979	–0.51053
Transparency (cm)	–0.1408	1	0.618277*	0.531535	0.555424*	–0.56287*	0.28786	0.327991	–0.41338
Total hardness (mg/l)	0.304704	0.618277*	1	0.54908*	0.399579	–0.28487	0.208146	0.143097	–0.36684
Alkalinity (mg/l)	0.203178	0.531535	0.54908*	1	0.1596	–0.8206	–0.25071	–0.41456	–0.16839
Dissolved oxygen (mg/l)	–0.21703	0.555424*	0.399579	0.1596	1	–0.156	0.088928	0.367781	–0.09043
pH	0.070068	–0.56287*	–0.28487	–0.8206**	–0.156	1	0.0284	0.236412	0.089988
BOD (mg/l)	0.171218	0.28786	0.208146	–0.25071	0.088928	0.0284	1	0.737451**	–0.50206
COD (mg/l)	–0.28979	0.327991	0.143097	–0.41456	0.367781	0.236412	0.737451*	1	–0.27488
Dissolved organic matter (mg/l)	–0.51053	–0.413380	–0.36684	–0.16839	–0.09043	0.089988	–0.50206	–0.27488	1
Nitrite nitrogen (mg/l)	–0.13797	0.34806	0.025048	–0.32353	0.299974	0.239903	0.452242	0.689993**	–0.04863
Nitrate nitrogen (mg/l)	–0.25306	0.552581*	0.078231	0.24357	0.32939	–0.40361	0.12478	0.387605	–0.05372
Ammonical nitrogen (mg/l)	–0.5077	–0.20534	–0.53636**	–0.70424**	0.330481	0.44111	0.134093	0.411888	0.410925
Total ortho-phosphate (mg/l)	0.635194*	0.10438	0.313774	–0.08327	0. 199977	0.147157	0.491991	0.248865	–0.43844
Phytoplankton (no./L)	–0.02819	–0.50516	–0.37938	–0.0986	–0.42286	0.179203	–0.39117	–0.60051*	0.603531*
Zooplankton (no./L)	0.257164	–0.32282	–0.14102	0.500679	–0.32454	–0.36082	–0.69987*	–0.75332**	0.100791
GPP (mgC/m³/h)	0.178897	–0.53429	–0.32191	0.224099	–0.26147	–0.16973	–0.55731*	–0.76294**	0.625021*
NPP (mgC/m³/h)	0.209341	–0.71313	–0.31702	0.005862	–0.27298	0.034746	–0.47199	–0.66471**	0.585801*

**: Significant at 1 per cent level; *: Significant at 5 per cent level.

Contd...

Table 20.2–Contd...

Parameters	*Nitrite Nitrogen*	*Nitrate Nitrogen*	*Ammoniacal Nitrogen*	*Total Ortho-phosphate*	*Phyto-plankton*	*Zooplankton*	*GPP*	*NPP*
Water temperature (°C)	–0.13797	–0.25306	–0.5077	0.635194**	–0.02819	0.257164	0.178897	0.209341
Transparency (cm)	0.348063	0.552581*	–0.20534	0.104389	–0.50516	–0.32282	–0.53429**	–0.71313**
Total hardness (mg/l)	0.025048	0.078231	–0.53636	0.313774	–0.37938	–0.14102	–0.32191	–0.31702
Alkalinity (mg/l)	–0.32353	0.24357	–0.70424**	–0.08327	–0.0986	0.500679	0.224099	0.005862
Dissolved oxygen (mg/l)	0.299974	0.32939	0.330481	0. 199977	–0.42286	–0.32454	–0.26147	–0.27298
pH	0.239903	–0.40361	0.44111	0.147157	0.179203	–0.36082	–0.16973	0.034746
BOD (mg/l)	0.452242	0.12478	0.134093	0.491991	–0.39117	–0.69987**	–0.55731**	–0.47199
COD (mg/l)	0.689993**	0.387605	0.411888	0.248865	–0.60051**	–0.75332**	–0.76294*	–0.66471**
Dissolved organic matter (mg/l)	–0.04863	–0.05372	0.410925	–0.43844	0.603531**	0.100791	0.625021**	0.585801**
Nitrite nitrogen (mg/l)	1	0.691613**	0.465877	0.477949	–0.28234	–0.62031*	–0.45871	–0.57946*
Nitrate nitrogen (mg/l)	0.691613**	1	0.117724	0.252147	–0.45685	–0.05593	–0.2659	–0.52332
Ammonical nitrogen (mg/l)	0.465877	0.117724	1	0.038206	0.06008	–0.54573*	–0.02214	–0.01988
Total ortho-phosphate (mg/l)	0.477949	0.252147	0.038206	1	–0.26011	–0.23651	–0.1569	–0.19563
Phytoplankton (no./L)	–0.28234	–0.45685	0.06008	–0.26011	1	0.157928	0.640893**	0.625848*
Zooplankton (no./L)	–0.62031*	–0.05593	–0.54573	–0.23651	0.157928	1	0.58878*	0.530756
GPP (mgC/m³/h)	–0.45871	–0.2659	–0.02214	–0.1569	0.640893**	0.58878*	1	0.900026**
NPP (mgC/m³/h)	–0.57946*	–0.52332	–0.01988	–0.19563	0.625848**	0.530756*	0.900026**	1

**: Significant at 1 per cent level; *: Significant at 5 per cent level.

Table 20.3: Coefficient of Correlation Between Different Physico-chemical and Biological Parameters (Seasonally)

Parameters	*Water Temperature*	*Turbidity*	*Total Hardness*	*Total Alkalinity*	*Dissolve Oxygen*	*pH*	*BOD*	*COD*	*DOM*
Water temperature (°C)	1.000	0.618	0.545	0.900*	–0.019	–0.642	–0.198	–0.477	–0.527
Transparency (cm)	0.618	1.000	0.717	0.427	0.700	–0.999**	0.646	0.393	–0.930*
Total hardness (mg/l)	0.545	0.717	1.000	0.689	0.717	–0.718	0.314	0.101	–0.907*
Alkalinity (mg/l)	0.900*	0.427	0.689	1.000	0.005	–0.451	–0.368	–0.609	–0.495
Dissolved oxygen (mg/l)	–0.019	0.700**	0.717	0.005	1.000	–0.680	0.859	0.764	–0.839
pH	–0.642	–0.999**	–0.718	–0.451	–0.680	1.000	–0.622	–0.364	0.926*
BOD (mg/l)	–0. 198	0.646*	0.314	–0.368	0.859	–0.622	1.000	0.956*	–0.620
COD (mg/l)	–0.477	0.393	0.101	–0.609	0.764	–0.364	0.956**	1.000	–0.388
Dissolved organic matter (mg/l)	–0.527	–0.930*	–0.907*	–0.495	–0.839	0.926*	–0.620	–0.388	1.000
Nitrite nitrogen (mg/l)	–0.292	0.490	–0.045	–0.578	0.594	–0.467	0.922*	0.923*	–0.344
Nitrate nitrogen (mg/l)	–0.096	0.430	–0.284	–0.502	0.237	–0.419	0.684	0.659	–0.146
Ammonical nitrogen (mg/l)	–0.781	–0.279	–0.690	–0.975**	0.009	0.301	0.449	0.649	0.419
Total ortho-phosphate (mg/l)	0.805	0.896*	0.473	0.516	0.314	–0.907*	0.358	0.087	–0.702
Phytoplankton (no./L)	–0.215	–0.888*	–0.461	0.035	–0.792	0.875*	–0.909*	–0.753	0.787
Zooplankton (no./L)	0.505	–0.334	–0.304	0.481	–0.866	0.304	–0.884*	–0.935*	0.462
GPP (mgC/m³/h)	0.423	–0.388	–0.423	0.366	–0.914*	0.359	–0.864	–0.891*	0.547
NPP (mgC/m³/h)	0.293	–0.545	–0.456	0.315	–0.947*	0.518	–0.943*	–0.925*	0.645

**: Significant at 1 per cent level; *: Significant at 5 per cent level.

Contd...

Table 20.3–Contd...

Parameters	Nitrite Nitrogen	Nitrate Nitrogen	Ammoniacal Nitrogen	Total Ortho-phosphate	Phyto-plankton	Zooplankton	GPP	NPP
Water temperature (°C)	–0.292	–0.096	–0.781	0.805	–0.215	0.505	0.423	0.293
Transparency (cm)	0.490	0.430	–0.279	0.896*	–0.888*	–0.334	–0.388	–0.545
Total hardness (mg/l)	–0.045	–0.284	–0.690	0.473	–0.461	–0.304	–0.423	–0.456
Alkalinity (mg/l)	–0.578	–0.502	–0.975**	0.516	0.035	0.481	0.366	0.315
Dissolved oxygen (mg/l)	0.594	0.237	0.009	0.314	–0.792	–0.866	–0.914*	–0.947*
pH	–0.467	–0.419	0.301	–0.907*	0.875*	0.304	0.359	0.518
BOD (mg/l)	0.922*	0.684	0.449	0.358	–0.909*	–0.884*	–0.864	–0.943*
COD (mg/l)	0.923*	0.659	0.649	0.087	–0.753	–0.935*	–0.891*	–0.925*
Dissolved organic matter (mg/l)	–0.344	–0.146	0.419	–0.702	0.787	0.462	0.547	0.645
Nitrite nitrogen (mg/l)	1.000	0.896*	0.695	0.331	–0.833	–0.732	–0.666	–0.765
Nitrate nitrogen (mg/l)	0.896*	1.000	0.678	0.478	–0.714	–0.353	–0.265	–0.410
Ammonical nitrogen (mg/l)	0.695	0.678	1.000	–0.321	–0.181	–0.455	–0.332	–0.327
Total ortho-phosphate (mg/l)	0.331	0.478	–0.321	1.000	–0.714	0.071	0.037	–0.156
Phytoplankton (no./L)	–0.833	–0.714	–0.181	–0.714	1.000	0.634	0.637	0.780
Zooplankton (no./L)	–0.732	–0.353	–0.455	0.071	0.634	1.000	0.991**	0.972**
GPP (mgC/m³/h)	–0.666	–0.265	–0.332	0.037	0.637	0.991**	1.000	0.979**
NPP (mgC/m³/h)	–0.765	–0.410	–0.327	–0.156	0.780	0.972**	0.979**	1.000

**: Significant at 1 per cent level; *: Significant at 5 per cent level.

reported to fall up to 2.7 ppm in summers during 1991-93 resulting into fish mortality (Marothia, *Op. cit.*). Such low levels of DO have not been recorded in this study. It shows seasonal and monthly variation throughout the year. The highest dissolved oxygen was recorded in the month of December and the lowest in the month of August. The higher DO content in winter is due to higher solubility of oxygen in water. The lower catabolic activity of the aquatic organisms probably accelerated the DO further in the water (Chourasia and Adoni, 1985 and Sukumaran and Das, 2002).

Alkalinity of water refers to the total concentration of bases present in water, mainly bicarbonates and carbonates and it is also an index of potential carbon dioxide (Chhattopadhya, 1995 and Lokare and Rathinaraj, 1997). Free CO_2 was always found to be nil and the bicarbonate alkalinity of water varied significantly over months and seasons. Alkalinity ranged from 84 mg/l to 147.5 mg/l in Budha Talab. Selot (*Op. cit.*) reported 120-180 mg/l alkalinity in Dulhara reservoir in Chhattishgarh. Kanungo and Naik (1987) studied nineteen-ponds of Raipur and found higher value of alkalinity (512.45 mg/l). The highest alkalinity was found in summer months and lowest in spring. According to Khatavkar and Trivedy (1993) the pollution of water can increase the level of total alkalinity of water but in afternoon due to the absence of carbon dioxide, it decreases the total alkalinity as found in this study. Higher alkalinity in summer is due to excess use of detergents and the entry of sewage water. A negative significant correlation was observed between pH and alkalinity.

Hardness of water refers to the total concentration of divalent cations (Calcium and Magnesium) in water (Lokare and Rathinraj, *Op. cit.*). It varied significantly over the months and seasons ranging between 103 mg/l to 179.6 mg/l. Hardness decreased in winter and increased in summer. It also showed positive correlation with temperature ($r = 0.304$). Similar result was also found by Chari (*Op. cit.*), Nandan and Mahajan (2000) and Bhatt *et al.* (1999). Excessive evaporation increases the concentration of salts and use of detergents also increases the hardness in waters. The lower hardness in monsoon may be due to dilution with the rainwater. Significant positive correlation was found with transparency ($r = 0.618$) and alkalinity ($r = 0.55$) *i.e.* hardness increased with an increase in transparency and alkalinity.

Ammonical nitrogen is a product of ammonification of organic matter (Ellis *et al.*, 1946; Rybak and Sikorska, 1976 and Barik *et al.*, 2005). The ammonical nitrogen in Budha Talab ranged from 1.55 to 4.80 mg/l in this study. It varied significantly over the months throughout the year. It was highest in the month of January and the lowest in the month of June. According to Ellis *et al.* (1946) organically polluted water body have ammonia greater than 1.0 mg/l. The present study also showed ammonia always above 1.0 mg/l. Chari (*Op. cit.*) and Basheer (1996) also found similar ammonical nitrogen values in city bound tanks. Ammonical nitrogen has a negative correlation with total alkalinity, hardness and temperature whereas a positive correlation was seen with COD, nitrite-nitrogen with respect to seasons. Higher temperature increases microbial growth and thereby faster decomposition produces more ammonia, which is in unstable form and in the presence of oxygen converted into nitrites and nitrates. Low ammonia in July may be attributed to dilution of water, which added nitrogen from outside sources and thereby upsets the natural cycle of ammonical and nitrate forms. (Mortimer and Hickling, 1959).

Nitrite-N is an intermediate product in the bacterial oxidation of ammonia to nitrate-N (Lokare and Rathinraj, *1997*). Nitrite nitrogen ranges between 0.02 to 0.17 mg/l and significantly varied over the months. The highest nitrite-N value was recorded in the month of February (0.17 mg/l) and the lowest was in the month of August. However it was not varied significantly over the seasons. Low amount of nitrite-N was also found by Khatavkar and Trivedy, (*Op. cit.*), and Kanungo and Naik (*Op.*

cit.). A positive significant correlation of nitrite was found with nitrate nitrogen (r = 0.691). Unionized form of nitrite-N become toxic in form of nitrous acid that can readily pass through the gill membrane than their ionized nitrite (NO_2^-). The toxic level of nitrite nitrogen is more than 0.2 mg/l. for aquatic animals. The concentration of nitrite nitrogen depends on pH and temperature (Barik, 2004). But in this study no significant correlation was found between nitrite-N, pH and temperature.

The nitrate nitrogen varied significantly over the months and seasons. The highest nitrate value was observed in the month of February (0.86 mg/l) and the lowest in the month of May (0.24 mg/l). It was also found that nitrate is positively correlated with BOD (r = 0.684), COD (r = 0.659), nitrite (r = 0.896) and ammonical-N (r = 0.678) and negatively correlated with phytoplankton. Wetzel (1975) also reported that, nitrate-N varied significantly over the seasons. According to Ganapati (1940) the non polluted tropical waters are generally deficient in nitrates but the factors like discharge of sewage, run off and nitrogen fixation increases the concentration in the water bodies as found in Budha Talab. Contribution of large human settlement and their discharge of domestic waste and sewage results into higher nitrate-N. Compared to ammonia and nitrite-N, nitrate is less toxic at similar concentration and readily soluble in water. Nitrate-N can readily be absorbed by autotrophs *i.e.* algae and bacteria.

Total orthophosphate varied significantly over the months. The highest orthophosphate value was recorded in the month of April (1.28 mg/l) and lowest was in December (0.23 mg/l). It also varied significantly over the season with the highest in summer (0.76 mg/l) and the lowest in winter (0.25 mg/l). Restricted water in summers results into higher values. Detergents are rich in phosphorus and they also contribute in restricted waters (Kanungo, 1986). Total orthophosphate showed positive correlation with temperature and transparency respectively (r = 0.805 and r= 0.896).

The highest DOM was recorded in the month of September (2.5 mg/l) and the lowest in the month of May (1.95 mg/l). There was no seasonal variation found in this study. As for season the highest DOM was found in monsoon (2.26 mg/l) and the lowest in summer (2.0 mg/l). A negative correlation was found between seasonal means of DOM and dissolved oxygen (r=–0.839). Low dissolved oxygen found in monsoon was due to high organic matter, which requires higher oxygen for decomposition. A negative significant correlation was seen with transparency ((r=–0.930). Low transparency in monsoon is due to inflow of domestic waste and sewage from near by settlements which increases the organic matter in Budha Talab. High temperature accelerates decomposition of organic matter (Chari, 1985). However in the present study no relation was found with temperature.

BOD is the amount of oxygen required to decompose organic matter by microorganisms in a water body. BOD varied significantly throughout the year ranging between 38.5 to 177.5 mg/l. The highest BOD was recorded in the month of March and the lowest in the month of July. Kanungo and Naik (1987) studied nineteen ponds of Raipur and found BOD ranges from 31.2 to 67.6 mg/l. Marothia (*Op. cit.*) reported that BOD value reaches upto 33 mg/l in Budha Talab. Whereas in this study BOD was found upto 177.5 mg/l due to higher population and increased inflow of sewage and organic matter. Higher BOD indicates greater organic pollution in water body (Chari, *Op. cit.*). A positive significant correlation was found with COD (r = 0.737), which also confirms pollution in the water body. Very high BOD was reported by Rao *et al.* (1978) upto 380 mg/l, Chattopadhyay *et al.* (1984) upto 460 mg/l and Khatavkar and Trivedy (*Op. cit.*) upto 422 mg/l. Relatively it may be said that Budha Talab has very less BOD (upto 177.5 mg/l).

COD varied from 71.0 to 266.5 mg/l in Budha Talab. The observed mean COD varied significantly over the months and seasons. The highest COD was recorded in the month of February and the lowest in the month of July (Table 20.1). Organic and inorganic chemical pollution in water is reflected by high value of COD. Khatavkar and Trivedy (*op. cit*) reported COD values to be more than 60 mg/l in organically polluted waters. Similar is the condition noticed in Budha Talab, where in COD was always higher than 71.0 mg/l. Low COD in monsoon may be attributed to dilution of water. A significant positive correlation of COD was found with BOD and nitrites (r = 0.737). Very high COD was reported by Govindan and Sundaresan (1979) upto 830 mg/l. and Kiran *et al.* (*Op. cit.*) upto 378 mg/l. On the other hand Rai (1978), Rao *et al.* (1978) and Raina *et al.* (1984) obtained values near about the present study.

References

APHA, 1998. *Standard Methods for Examination of Water and Wastewater*, 17th edn. American Public Health Association, Washington, DC, pp. 1525.

Barik, P., Vardia, H.K. and Gupta, S.B., 2005. Exploiting nitrifying bacterical isolates (*Nitrosomonus and Nitrobacter*) as a bioremediator of ammonia and nitrite in fish ponds of rice ecosystem. *J. Agril. Issue*, 10(2): 49–52.

Barik, P., 2004. Bioremediation of ammonia in aquaculture system by the use of nitrifying bacteria (*Nitrosomonas* sp. and *Nitrobacter* sp.). *M.Sc. Thesis*, Indira Gandhi Agricultural University, Raipur. p. 22.

Basheer, V.S., Asif, A. and Aftab, A., 1996. Seasonal variation in the primary productivity of a pond receiving sewage effluents. *Journal of Inland Fishery. Soc., India*, 28(1): 76–82.

Bhatt, L.R., Laucoul, P.H.D. and Jha, P.K., 1999. Physico-chemical characteristics and phytoplankton of Taudaha lake, Kathmandu. *Poll. Res.*, 18(4): 353–358.

Chari, M.S., 1985. Aquatic pollution and its effect on the fauna and flora of a freshwater pond at Aligarh, India. *Geobios* (Spl. Vol. Proc. Nat. Sympos. Evalu. Env. Ends.), p. 49–65.

Chattopadhya, N., 1995. *Nutrient Management in Aquaculture*. Soil Testing Laboratory Institute of Agriculture, West Bengal, p. 35.

Chattopadhyay, S.N., Routh, Tapan, Sharma, V.P., Arora, H.C. and Gupta, R.K., 1984. A short term study on the pollutional status of river Ganga in Kanpur region. *Indian J. Environ. Hlth.*, 26: 244–257.

Chourasia, S.K. and Adoni, A.D., 1985. Zooplankton dynamics in a shallow eutropical lake. In: *Proc. Nat. Symp. Pure and Appl. Limnology, Bull. Bot. Soc.*, 32: 30–39.

Ellis, M.M., Westfall, B.A. and Ellis, M.D., 1946. Determination of water quality. Fish and Wildlife Service. *U.S. Department Intevoir Res. Report*, 9. 122.

Ganapati, S.V., 1940. The ecology of a temple tank containing a permanent bloom of *Microcystis aeruginosa* (Kutz.) Henfr. *J. Bombay Nat. Hist. Soc.*, 42: 65–77.

Golterman, H.L., 1970. *Methods for Chemical Analysis for Freshwater*. IBP Handbook No. 8 Blackwell Scientific Publications, Oxford and Edinburgh, p. 156–187.

Govindan, V.S. and Sundaresan, B.B., 1979. Seasonal succession of algal flora in polluted region of the Adyar river, India. *Indian J. Environ. Hlth.*, 21: 131–142.

Hutchinson, G.E., 1957. *A Treatise on Limnology*. John Wiley and Sons Inc., 1: 57–59.

Kanungo, V.K., 1986. Ecology of public wastewater of Raipur city. *Ph.D. Thesis*, Ravishankar University, Raipur, pp. 1–322.

Kanungo, V.K. and Naik, M.L., 1987. Physio-chemical and biological characteristics of nineteen ponds of Raipur. In: *Perspective in Hydrobiology*, 30: 157–160.

Khatavkar, S.D. and Trivedy, R.K., 1993. Ecology of freshwater reservoir in Maharashtra with reference to pollution, physico-chemical characteristics of water bodies and trophic status. In: *Ecology and Pollution of Indian Lakes and Reservoirs*, (Eds.) P.C. Mishra and R.K. Trivedi. Ashish Publishing House, New Delhi. pp. 25–56/57–83.

Kiran, R., Deepa, R.S. and Ramachandra, T.V., 1998. Comparative water quality assessment of Bannerghatta and yediyur lakes of Bangalore. In: *Proceeding of the National Seminar on Environmental Pollution: Cause and Remedies*, pp. 166–182.

Lokare, K.V. and Rathinraj, G., 1997. *Aquaculture Engineering and Water Quality Management*. MPEDA, Kochi, pp. 45–63.

Marothia, D.K., 1997. Property rights, externalities and environmental pollution. In: *The Challenge of the Balance*, (Ed.) Anil Agrawal. Center for Science and Environment, New Delhi, pp. 290–298.

Mohanty, R.K., 1999. Effect of water Transparency on the growth of black tiger shrimp, *Penaeus monodon*. *J. Aqua.*, 7: 25–32.

Mortimer, C.H. and Hickling, C.F., 1954. Fertilizers in fish pond. *Fish. Pub. Lond.*, 5: 155.

Mukhopadhya, P.K. and Sarangi, N., 2006. Wastewater re-use and nutrient recycling system through aquaculture. *Fishing Chimes*, 25(10): 74–78.

Nandan, S.N. and Mahajan, S.R., 2000. Pollution status of Hartala Lake of Jalgaon district (Maharashtra). In: *Ecology of Polluted Water*. APH Publishing, New Delhi, 2: 333–350.

Rai, L.C., 1978. Ecological studies of algal communities of the Ganges river at Varanasi. *Indian. J. Ecol.*, 5: 1–6.

Raina, V., Shah, A.R. and Shakti, R. Ahmed, 1984. Pollution studies on river Jhelum. In: An assessment of water quality. *Indian. J. Environ. Hlth.*, 26: 187–201.

Rao, S.V.R., Singh, V.P. and Mall, L.P., 1978. Biological methods for monitoring water pollution levels studied at Ujjain. In: *Glimpses of Ecology*, (Eds.) J.S. Singh and B. Gopal. Commemoration Volume International Scientific publication, Jaipur, India, pp. 341–348.

Ruttner, F., 1963. *Fundamentals of Limnology*, Transl. D.G. Frey and F.E.J. Fry. University of Toronto Press, Toronto, pp. 307.

Rybak, J.J. and Sikorska, V., 1976. Environment. In: *Selected Problems of Lakes Littoral Ecology*, (Ed.) E. Pieczynska. Univ. of Warsaw, p. 113.

Selot, M., 1977. Ecological study of Dulhara tank with special reference to plankton population and physico-chemical parameters of water. *Ph.D. Thesis*, Pt. Ravi Shankar Shukla University, Raipur. p. 196.

Sreenivasan, A., 1969. Eutrophication trends in a chain in Madras (India). *Env. Hlth.*, 11: 392–401.

Sukumaran, P.K. and Das, A.K., 2002. Plankton abundance in relation to physico-chemical features in a peninsular man made lake. *Environment and Ecology*, 20(3): 873–879.

Wetzel, R.G., 1975. *Limnology*. W.B. Saunders Co., Philadelphia, p. 743.

Zutshi, D.P. and Vass, K.K., 1984. Limnological studies on Dal lake Kashmir. V. Impact of human activities and the evolution of lake environment. In: *Paper Presented at National Seminar on Environment*, Bhopal, Feb. pp. 8–10.

Chapter 21

Studies on Algal Flora and Physico-chemical Characteristics of Shikara Reservoir in Nanded District, Maharashtra

☆ *S.D. Dhavle, H.M. Lakde and S.D. Lohare*

ABSTRACT

Algal flora is an important component in aquatic food chain. The present investigation deals with the study of algal flora and water characteristics from Shikara, reservoir. The seasonal variation in water temperature, pH, dissolved oxygen, free CO_2, alkalinity, chlorinity, total hardness, total solids, bicarbonates, nitrates, phosphate and calcium were studied. Algal flora is dominated by Chlorophyceae followed by Bacillariophyceae and Cyanophyceae.

Introduction

Shikara, a minor reservoir is situated near Mukhed in Nanded district of Maharashtra. The area water spread area is 81 hectares and hight of reservoir is about 47 feet. Zafar (1959) reported periodicity of unicellular algae from waterbodies of Hyderabad. The study of aquatic environment and ecology of phytoplankton in freshwater studied by Jaybhaye *et al.* (2007).

The present investigation deals with the study of algal flora and seasonal variation in physico-chemical parameters of Shikara reservoir for the period of January 2007 to December 2007. The reservoir water is mainly used for irrigation, drinking, fishery and domestic purposes.

Materials and Methods

Algal samples were collected during January 2007 to December 2007 and analysed by standard methods (APHA, 1975). Algae were indentified following the relevant monograph (Hustadt, 1930; Pochman, 1942; Desikachary, 1959). The samples were collected airtight and opaque polythene containers.

The analysis of physico-chemical characteristics was carried out by standard methodology for water analysis given by Kodarkar *et al.* (1998) and Trivedy and Goel (1986).

Table 21.1: Algal flora of Shikara Reservior

1.	Chlorophyceae	*Pandoring norum (Mull), Scenedesmus quadricauda (Turp), Pediastrum simplex Meyen., Spirogyra Karnalae, Chara toetida, Oedogonium Crassum, Cosmarium reniformae* and *Chlorococcum humicola (Nag.)*
2.	Bacillariophyceae	*Navicula radiosa kutz, Nitzchia fasciculata Grun, Pinnularia viridis, Cymbella laceolata* and *Fregillaria brevistriate mayer.*
3.	Cyanophyceae	*Oscillatoria subbrevis, Rivularia, mehrai* and *Spirulina meneghinia ana.*

Results and Discussion

In the present investigation three groups of algae *viz.* Chlorophyceae, Bacillariophyceae and Cyanophyceae comprising 16 species were recorded (Table 21.1). Green algae and diatoms were found to be dominant. The values of physico-chemical parameters are shown in Table 21.2. There was no any significant difference in the range of physico-chemical parameters in rainy season.

Table 21.2: Physico-chemical Properties of Shikara Reservoir

Sl.No.	*Parameters*	*Range*
1..	Water temperature (°C)	24°C–30°C
2.	pH	7.2–8.7
3.	Dissolved Oxygen (mg/l)	6.1–8.5
4.	Free CO_2 (mg/l)	0.0–0.32
5.	Alkalinity (mg/l)	40-90
6.	Total hardness (mg/l)	90-120
7.	Chlorinity (mg/l)	20-26
8.	Total Solids (mg/l)	190-540
9.	Carbonates (mg/l)	40-90
10.	Nitrates (mg/l)	0.85–1.30
11.	Phosphates (mg/l)	0.06–1.5
12.	Calcium (mg/l)	3.8–29.8

Temperature is most important factor affecting the growth of diatoms (Patric *et al.,* 1968). In present study during high temp the growth of diatoms was maximum (Nandan S.N. 1993). At high conc. of pH showed significant effect of population of blue green algae and dissolved oxygen in water is favourable for higher percentage of Chlorophyceae. The other factors like alkalinity, chlorinity, hardness, solids, carbonates, nitrates and calcium influenced the growth and development of algal flora.

References

APHA, 1975. American Publication Health Association, New York, 680.

Desikachary, T.V., 1959. *Cyanophyta.* ICAR, New Delhi, pp. 686.

Hustadt, F., 1930. *Bacillariophyta (Diatomaceae).* 10 P.A. 466.

Jaybhaye, V.M., Madlapure, V.R. and Salve, B.S., 2007. *Maharashtra J. Aqua. Biol.,* 22(2): 27–32.

Kodarkar, M.S., Diwan, A.D., Murugan, N., Kulkarni, K.M. and Ramesh, Anuradha, 1998. Methodology of water analysis: Physico-chemical, biological and microbiological. *Indian Association of Aquatic Biologists*, Hyderabad, pp. 12–102.

Nandan, S.N., 1993. Algal flora of fish pond in Dhule, Maharashtra. *I.B.R.*, (1 and 2): 61–63.

Trivedy, R.K. and Goel, P.K., 1986. *Chemical and Biological Methods for Water Pollution Studies*. Env. Pub., Karad.

Zafar, A.R., 1959. *Indian Bot Soc.*, 38: 549.

Chapter 22

On a New *Senga waranensis* (Cestoda : Ptychobothridae) from *Mastacembellus aramatus* at Warnanagar in Maharashtra

☆ *L.P. Lanka, S.R. Patil, A.D. Mohekar and B.V. Jadhav*

Introduction

The genus Senga was established by Dollfus (1934) with its type species, *S. besnardi* from *Betta splendens*. The Siamese fighting fish in an aquarium at Vinecunes, France, *S. ophiocepalina*. Tseng (1933) as *Anchistrocephalus ophiocephalina* from *Ophiocephalus argus* at Taimen, China and identified with a form previously recorded by Southwell (1913) as *Anchitrocephalus polyptera* (Anchitrocephalus) Monticelli 1890. Syn. Anchitrocephalus Luhe (1999), from *Ophiocephalus striatus;* in Bengal, India. *S. pycnomera* from *Ophiocephalus marulius* at Allahabad, India. *S. lucknowensis*. Johri (1956) from Mastacembellus in India, Fernando and Furtado (1983) recorded *S. malayana* from *Channa striata*, *S. parva* and *S. filiformis* from *Channa micropeletes* at Malacca.

Ramadevi and Hanumantrao (1966) reported the plerocercoid of *Senga* sp. From *Panchax panchax*, Tadros (1868) synomised the genus Senga with the genus Polyonchobothrium and proposed new comb for the species, Furtado and Chauhan (1971) reported *S. pahangensis* from *Channa micropeltes* at Tesak Bera, Shinde (1972) redescribed *S. besnardi* from *Ophiocephalus gachua* in India and recently Ramadevi and Rao (1973) reported another species *S. vishakapatanamensis*, from *Ophicephalus punctatus* at Vishakapatnam, India. Ramadevi (1976) described the life cycle of *S. vishakapatanamensis* from *Ophiocephalus punctatus* in a lake at Kondakaria, Andhra Pradesh, but they do not agree with the Tadors statements, Wardle, McLeod and Radinovsky (1974) put Senga as a distinct genus in the family ptychobothridae, Deshmukh (1980) reported *S. khami* from *Ophiocephalus marulius* a freshwater fish from Kham river at Aurangabad. Jadhav and Shinde (1980) added new species *S. aurangabadensis* from *Mastacembellus armatus* at Aurangabad. Then Shinde and Jadhav (1980) described a new species *S. godavari* from *Mastacembellus armatus*. A new species was added by Kadam *et al.* (1981) as *S. paithaniensis* from *Mastacembellus armatus*. The new species is added to genus by *Jadhav et al.* (1991) as *S. maharashtri* from the intestine of *Mastacembellus armatus* at Daryapur, Amravati. Tat and Jadhav (1997) added new species *S. mohekarae* from the intestine of the *Mastacembellus arnatus* at Pali in Beed district of Maharashtra. Later on no species is added to this genus. The present form is collected from *Masatcembellus armatus* at Warnanagar in Kolhapur district of Maharashtra in month of November 1999.

Materials and Methods

About ten cestodes were collected from intestine of freshwater fish *Mastacembellus armatus* at Warnanagar in the month of November 1999. They were preserved in 4 per cent formalin, stained with Harris haematoxylin, dehydrated through alcoholic grades and mounted in D.P.X. The figures are drawn with the help of camera Lucida. All measurements are in millimeters.

Senga waranensis n.sp.

Description: Ten specimens of cestode parasites were collected from a freshwater fish *Mastacembellus armatus* (Lacepede). The worms were flattened, consisting of scolex, immature, mature and gravid segments. The scolex is triangular, anterior end is pointed and posterior end is broad and measures 1.378 to 1.478 in length and 0.348 to 0.947 in breadth. The scolex bears botheria, which extends from the anterior end to posterior end of the scolex. The botheria measure 1.061 to 1.350 in length and 0.017 to 0.156 in breadth. The anterior end of scolex terminates into a rostellum,which is armed around to oval in shape and measures 0.265-0.303 in length and 0.098 to 0.152 in width. The rostellum is armed with circularly arranged hooks and are 42-45 in numbers. The hooks are stout, single pronged of unequal length, pointed at the apex, shorter hooks are present in the centre, longer hooks measures 0.035 to 0.087 in length and 0.022 to 0.043 in width. The shorter hooks measures 0.029 to 0.333 in length and 0.019 to 0.017 in width.

The neck is long, tapering towards anterior side, wide and measures 0.469 to 0.477 in length and 0.323 to 0.341 in breadth.

The mature segments are broader than long, about three times broader than long and measures 0.644 to 0.674 in length and 1.652 to 1.841 in breadth. The testes are small, rounded in shape, 250-260 in number, distributed density at the lateral side of the segment, on either side of the ovary and cirrus pouch and measures 0.197 to 0.098 in diameter.

The cirrus pouch is oval in shape, obliquely placed pre-ovarian in position, situated the center half of the segment. It measures 0.192 to 0.197 in length and 0.082 to 0.098 in breadth. It opens at its distal end by common genital opening at the middle of the segment. The cirrus is long, thin and extends beyond the cirrus pouch and measures about 0.806 to 0.811 in length and 0.023 in breadth. The vas deferens is short, thin and coiled and extends anteriorly and measures 0.043 to 0.045 in length and 0.008 to 0.009 in breadth. The genital pore is small in size, oval in shape and measures 0.027 to 0.030 in length and 0.022 to 0.023 in breadth.

The ovary is large in size, distinctly bilobed with unequal margins and transversely placed near the posterior margin of the segment, situated in the centre of the segment and measures 0.756 to 0.781 in length and 0.031 to 0.038 in breadth.

The vagina is a thin tube, slightly coiled, arises from the genital pore runs posteriorly obliquely and open into the ootype and measures 0.023 to 0.029 in length and 0.006 to 0.008 in breadth. The ootype is rounded to oval, medium in size, present between the ovarian lobes and measures 0.762 to 0.781 in length and 0.031 to 0.038 in breadth.

The vitellaria are granular each lateral side from anterior to posterior margin of the segment, arranged in 3-4 rows on lateral side.

The gravid segments shows saccular uterus placed at the anterior region of the segment, opens through a thick walled, double layered pore and dorsal in position, which measures 0.893 to 0.902 in length and 0.128 to 0.136 in breadth. Uterus is filled with eggs, eggs are oval in shape, yellowish in colour with some blackish border and measures 0.250 to 0.333 in length and 0.091 to 0.129 in breadth.Uteruine pore is rounded, towards anterior region of the segment.

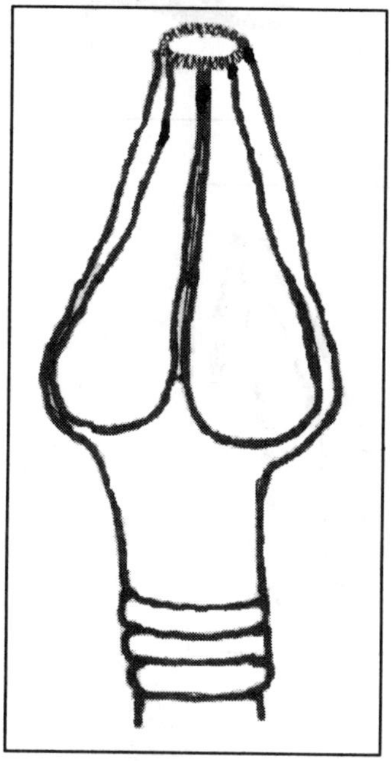

Scolex

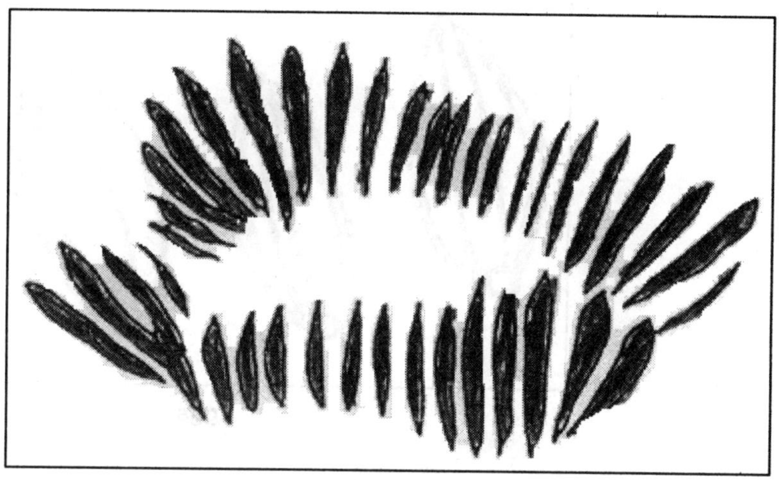

Hooks

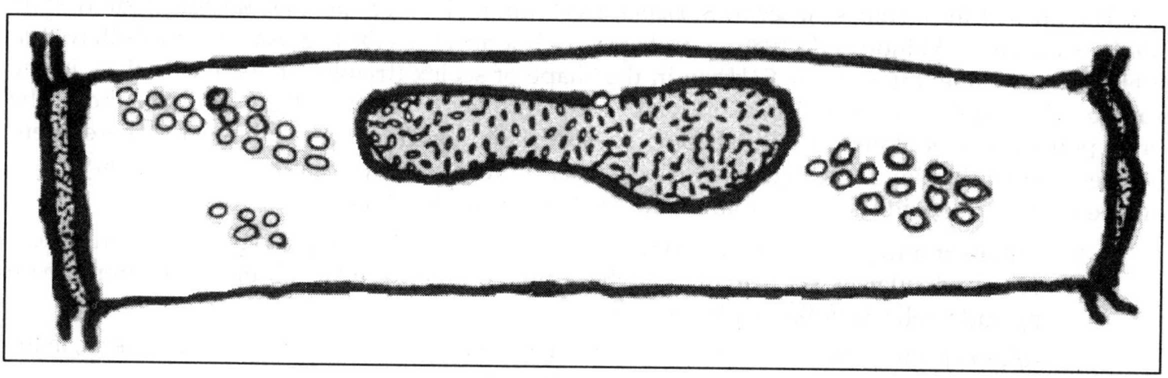

Gravid Segment

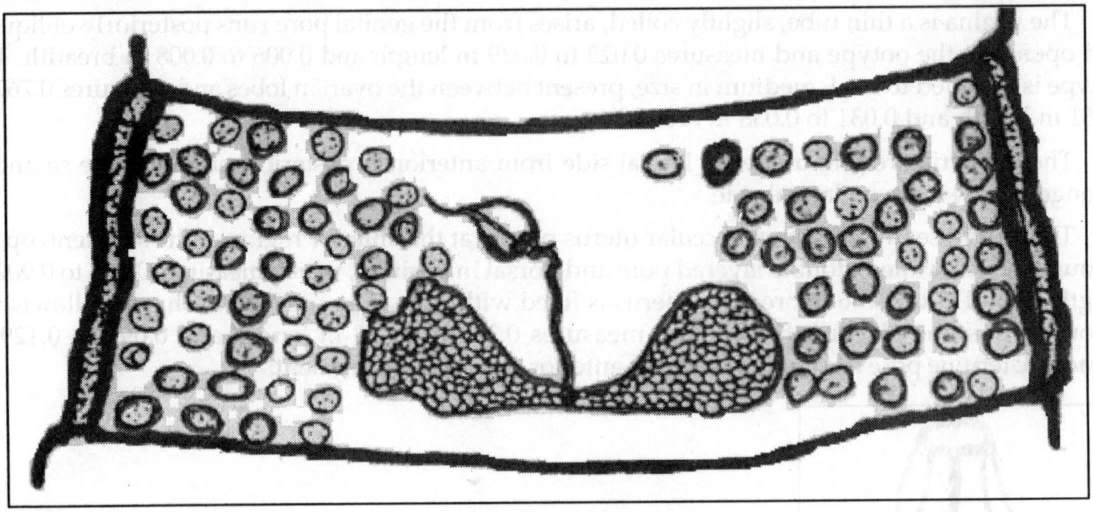

Mature Segment

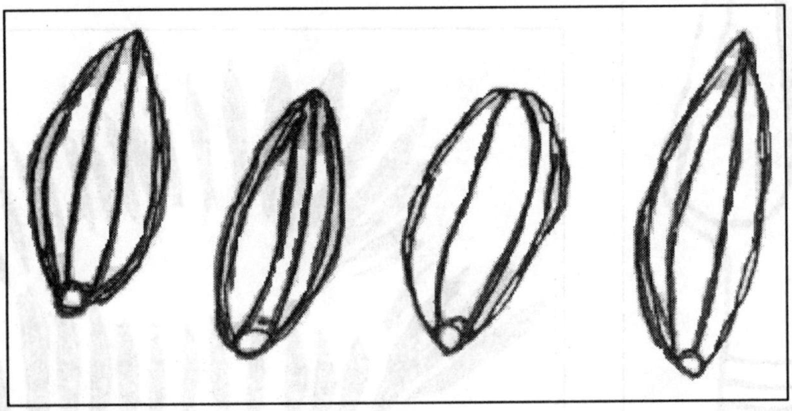

Eggs

Discussion

The present form comes closer to *S. khami*, Deshmukh (1980), *S. aurangabadensis*, Jadhav *et al.* (1980), S. Godavari, Shinde *et al.* (1980), *S. paithanensis*, Kadam *et al.* (1981), *S. maharahtrii*, Jadhav *et al.* (1991) but the same differs from *S. khami* in the shape of scolex (triangular Vs. rectanular), in the number of hooks (42-44 Vs. 55-57), short neck, in the number of testes (250-260 Vs. 155) and in the cirrus pouch (oval Vs. elongated), the shape of scolex (triangular Vs. oval) size of the mature segment (3 times broader than long), arrangement of testes (distributed densely on either side the iovary Vs. in two field) and arrangement of follicular vitellaria (4–5 in rows Vs. 2-3 rows).

1. It differs from *S.godavarii* in the shape of the scolex (triangular Vs. pear shaped), arrangement of hook (circular Vs. semicircular), in the ootype (round Vs. oval), absence of operculum on egg and vitellaria (follicular Vs. granular).
2. It differs from *S. paithanensis* in the number of rostellar hooks (42-44 Vs. 54), short neck, in the number of testes (250-260 Vs. 130-135), position of vagina (anterior Vs. posterior).

3. It differs from *S.maharashtrii* in the shape of scolex (triangular Vs. oval), in the number of testes and in the shape (250-260 rounded Vs. 80–90 oval), position of genital pore (in the anterior half of the segment Vs. in the posterior half of the segment).
4. The present cestode differs from *S. mohekarae* in the shape of scolex (triangular Vs. pear shaped, broad in middle), size of mature segment (3 times broader than long Vs. 4-5 times broader than long), arrangement of testes (distributed densely on either side of ovary Vs. in two fields), number of hooks (42-44 Vs. 22-26), number of testes (250-260 Vs. 60-70).

As the differentiating characters are extinct, so here is reported a new species as *S.waranensis n.sp.*, collected from Warana. It is named after the locality Warana.

Acknowledgements

The authors are thankful to Head, Department of Zoology, Dr. Babasaheb Ambedkar Marathwada Univeristy, Aurangabad and Principal, Vivekanand College, Kolhapur for providing laboratory facilities.

References

Chincholikar, L.N. and Shinde, G.B., 1977. Studies on Indian cestodes redescription of *Senga ophiocephaline* Terng, 1933. *Ibid*, 16: 181–183.

Deshmukh, A. and Shinde, G.B., 1980. On a new cestode *Senga khami* n. sp. (Cestoda–Ptychobothridae) from freshwater fish. *Indian J. of Helminth.*, 31(1–2): 28–32.

Gairola, D. and Malhotra, S.K., 1986. Cestode fauna of fishes in river Ganges around an Indian sub-humid region. I. *Senga gangesii n. sp.* from *Mystus vittatus. Japanese. J. of Parasite*, 35(6): 471–474 (En. Ja. 13 ref). Parasit, Lab. Dept. of Zoology, University, Allahabad, U.P. , India.

Gairola, D. and Malhotra, S.K., 1987. Cestode fauna of food fishes in a sub-humid region of India. II. *Senga vittati n. sp.* from *Mystus vittatus. Acta. Parasitologia. Lituanica.* 22: 93–96, Ru. Li. En. Ref. Fac. Zool. Allahabad University, Allahabad, U.P., India.

Hiware, C.J., 1999. On a new tape worm *Senga armatusae* n. sp. from freshwater fish, *Mastacembelus armatus* at Pune, M.S. India. *Rivista Di. Parasit*, 16(60)N1: 9–12.

Johri, L.N., 1956. A new cestode *Senga lucknowensis* from *Mastacembellus armatus. Current Science*, 25(6): 193–195.

Jadhav, B.V. and Shinde, G.B., 1980. A new tape worm *Senga godawari* n. sp. from *Mastacembellus* at Aurangabad. *Indian Biology.*, 2(4): 46–48.

Jadhav, B.V., 1981. *Sena paithensis* n. sp. (Cestoda : Ptychocothridae) from *Mastacembellus armatus. Bioresearch*, 5(1): 95–96.

Jadhav, B.V., 1991. A new tapeworm *Senga gachua* n. sp. from the fish *Channa gachua* at Aurangabad. *Ibid*, 3(1): 39–41.

Kadam, S.S. , Jadhav, B.V. and Shinde, G.B., 1981. On a new cestode *Senga paithanesis n. sp.* (Cestoda : Ptychobothridae) from *Mastacembellus armatus. Bioresearch*, (1): 95–96.

Majid, M.A. and Shinde, G.B., 1984. Two new species of the genus *Senga Dollfus*, 1934 (Cestoda : Pseudophyllidea from freshwater fish at Jagannathpuri, Orissa. *Ind. J. of Parasitology*, (1): 169–172 (En. 10 Ref.) Dept. of Zool. Poona College, Poona, India.

Ramadevi, P. and Rao, K.H., 1966. Pleucrcoid of *Senga* sp. (Pseudophyllidea : Ptychobothridae) from the freshwater fish, *Panchax panchax* (Ham and Buch.). *Current Science*, 35(24): 626–627.

Ramadevi, P., 1973. On *Senga vishakapatnamensis* n. sp. (Cestoda : Preudophyllidea) from the intestine of freshwater fish, *Ophiocephalus punctatus* (Bloch.). *Riv. Parasit.*, 84: 281–286.

Shinde, G.B. and Jadhav, B.V., 1980. A new tape worm *Senga godavarii* n. sp. from *Mastacembellus armatus* at Aurangabad, India. *Ibid*, 11(410): 40–48.

Shinde, G.B. and Jadhav, B.V., 1980. A new tapeworm *Senga godavri* at Aurangabad, India. *Biology*, 1(4): 46–48.

Shinde, G.B. and Deshmukh, R.A., 1980. On *Senga khami* (Cestoda : Ptychobothridae) from the freshwater fish. *Indian J. Zoology*, 8(1–2).

Tat, M.B. and Jadhav, B.V., 1997. *Senga mohekarae n. sp.* (Cestoda : Ptychobothridae) from *Mastacembellus armatus*. *Rivista di Parasitologia*, 17(57)2: 293–296.

Chapter 23

On a New *Hexacanalis trygoni* (Cestoda : Lecanisephalidae) from *Trygon zugei* at Malvan in Sindhudurg District of Maharashtra

☆ *L.P. Lanka and B.V. Jadhav*

Introduction

The genus *Hexacanalis perrenoud* (1931) was recorded by Southwell (1911), as cephalobothrium with its type species *C. abruptum* from *Pteropatea micrura* and *Dasybatus khuli* at Ceylon (trivandrum coast) and he also reported another species *C.variabilis* from *Pristis cuspidatus* and *Dasybatus khuli* at Ceylon. Perrenoud (1931) synonymies these species and created a new genus Hexacanalis with its type species *H. abruptus* and the name *H. variabilis* was proposed for *C. variabilis*. He recorded the genus from the same locality (Ceylon), Shinde and Deshmukh (1979) reported two new species *H. zugei* and *H. yamaguti* from *Trygon zugei* and *Dicerobatis eregodoo* from Ratnagiri and Veraval (west coast of India) respectively. Shinde *et al.* (1982) added two new species *i.e., H. Trygoni* and *H. ratnagirensis* from *Trygon zugei*. Later on Murlidhar (1986) added one new species *i.e., H. indiraji* from *Trygon sephen* at Kakinada (Andhra Pradesh), East coast of India. In 1991, Shinde *et al.,* added *H. thapari* from *Trygon zugei* at Ratnagiri. Later on no species is added to this genus.

The present communication deals with the description of a new species *Hexacanalis trygoni* collected at Malvan in Maharashtra state in the month of January 2000.

Materials and Methods

About ten cestodes were collected from the intestine of marine water fish *Trygon zugei* at Malvan in Kolhapur district of Maharashtra in the month of January 2000. They were preserved in 4 per cent formalin, stained with Harris haematoxylin, dehydrated through alcoholic grades and mounted in D.P.X. The figures are drawn with the help of camera Lucida. All measurements are in millimeters.

Description

The specimens of the cestode were obtained from the spiral valve of *Trygon zugei* (Muller and Henle) at Malvan, west coast of India, in the month of January 2000. The worm measures about 2.086 mm in length.

The scolex is almost quadrangular compressed on all sides, divided in two, anterior and posterior measures 0.092-0.118 mm in length and 0.109-0.166 in breadth, anterior region highly muscular occupied by large protrusible oval suckers without central transverse slit. The posterior region bears four oval suckers placed at four corners, measurees 0.049-0.072 in length and 0.039-0.0.049 in breadth. Scolex followed by young segments and mature segments and gravid segment.

Mature segments broader than long almost two times broader than long, but last segment is elongated lateral borders convex, measures 0.092-00118 in length and 0.109-0.166 in breadth. Testes mature segments broader than long, almost two times broader than long but last segment is elongated lateral borders convex, measures 0.092-0.0118 in length and 0.109-0.166 in breadth. Testes 25-30 in numbers, rounded to oval, irregularly scattered in the centre of the segment it measures 0.024-0.029 in length and 0.010-0.015 in breadth, cirrus pouch to the lateral margins and alternate in arrangements, cirrus pouch is conical, elongated placed just anterior to the segment, opens submarginally and measures 0.053-0.072 in length and 0.015-0.043 in breadth. The cirrus is wide, little oiled and measures 0.072-0.087 in length and 0.005-0.005 in breadth. The genital pore irregularly alternate, measure 0.005 in length and 0.001-0.002 in breadth.

The ovary is a single mass, almost dup-bell shaped with acini enter of the segment and measures 0.057-0.066 in length and 0.007-0.019 in breadth.

The vagina is a thin tube, posterior to cirrus pouch, starts from genital pore, takes a curve runs posterior in the middle of the segment, reaches and opens into ootype and measures 0.019-0.024 in

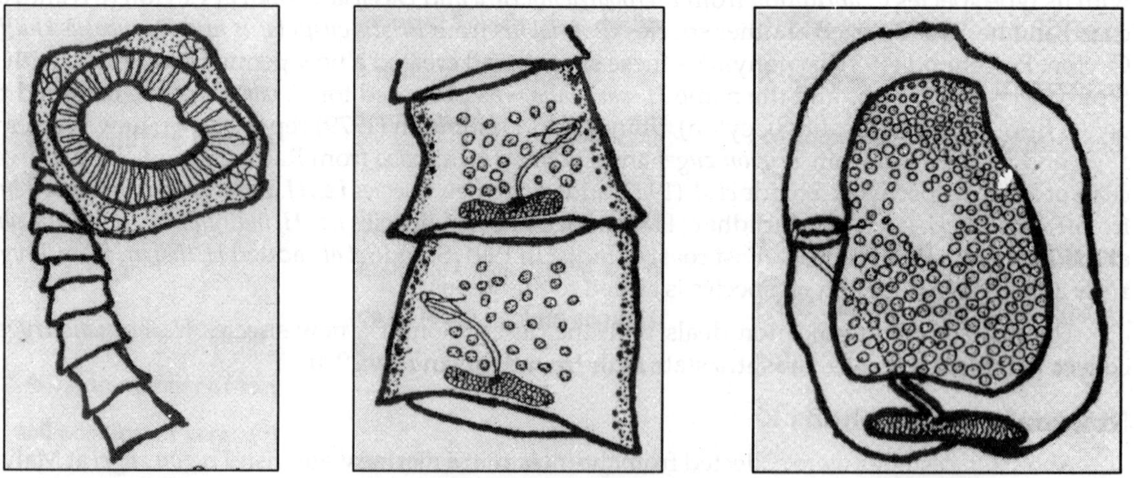

Scolex **Mature Segment** **Gravid**

Figure 23.1: ***Hexacanalis trygoni*** **sp.nov.**

length and 0.017-0.018 in breadth, vitellaria granular, corticular, wide strip from anterior to posterior margin of the segment.

Gravid segments are comparatively large, oval in shape, it measures about 0.492-0.505 in length and 0.467-0.489 in breadth, segments is filled with round to oval shaped eggs number 85-90.

Discussions

1. The worm under discussion comes closer to *H.abruptus*, Perranoud (1931), in the presence of receptaculum seminis but for some difference with the characters the central suckers (without transverse slit like opening Vs. with transverse slit like opening), presence of neck, in the number of testes (90-98 Vs. 60-63), in the shape of the ovary (quadrangular Vs. dumbell shaped) arrangement of vagina, position of genital pore (marginal Vs. submarginal) and the shape of vitellaria (granular Vs. follicular).
2. The tapeworm differs from *H.variabilis*, Perrenoud (1931), in the number of testes (90-98 Vs. 40-60 in number) in the shape of ovary (quadrangular Vs. elonagted)position of vagina (posterior Vs. anterior to cirrus pouch), position of genital pore (marginal Vs. submarginal) and the shape of vitellaria (granular Vs. follicular).
3. *H.trygoni n.sp.*comes closer to *H.zugei*, Shinde and Deshmukh (1979) in the presence of neck but the same differs from the number of testes (25-30 Vs. 42-47 in number), in the shape of ovary (quadrangular Vs. oval), presence of receptaculum seminis and position of genital pore (marginal Vs. submarginal).
4. The presence cestode differs from *H.yamaguti*, Shinde and Deshmukh (1979) in the presence of neck, in number of testes (90-98 Vs. 40-45 in number), shape of ovary (quadrangular Vs. elongated) and position of genital pore (marginal Vs. submarginals).
5. The present tapeworm comes close to *H.trygoni*, Shinde *et al.* (1982) in the presence of neck, where as it differs further from the same in the number of testes (90-98 Vs. 40-45 in number), in the shape of ovary (quadrangular Vs. compact) presence of receptaculum seminis, position of genital pore (marginal Vs. submarginal) and arrangement vitellaria (corticular and subcorticular).
6. The presence form differs from *H.rtangiriensis*, Shinde *et al.* (1982) with the characters *i.e.*, of central sucker isthmus transverse slit like opening, in the number of testes (90-98 Vs. 35), presence of receptaculum seminis and position of genital pore (marginal Vs. submarginal).
7. The present tapeworm comes close to *H.indiraji*, Murlidhar (1986) in the presence of neck but the same differs in the number of testes (90-98 Vs. 35-40), in the shape of ovary (quadrangular Vs. triangular) arrangement of vagina (posterior Vs. antrior to cirrus pouch), presence of receptaculum seminis and position of genital pore (marginal Vs. submarginal).
8. The present form differs from *H.thapari*, Shinde *et al.* (1991) with the characters, scolex quadrangular, anterior region small and posterior region large, long neck, mature segments broader than long, testes 60-65 in number, cirrus pouch cylindrical, genital pores marginally, cirrus coiled and granular vitellaria.

Hence keeping in view the different morphological characters which have been also inserted in the table and an insignificant at different levels and position, the authors proposes as new species *Hexacanalis tryogoni n.sp.* named after the host *Trygon*.

Acknowledgements

The authors are thankful to Head, Department of Zoology, Dr.Babasaheb Ambekar Marathwada University, Aurangabad and Dr.A.N.Jagtap, Principal, Vivekanand College, Kolhapur for providing laboratory facilities.

References

Murlidhar, A., 1986. A new species *Heaxacanalis indiraji sp*.n. (Lecanicephallidae) from a marine fish *Trygon sephen* at Kakinada (East coast of India). *Indian Journal of Helminthology*, 38(2): 151–156.

Shinde, G.B. and Deshmukh, R.A., 1979. On two new species of Hexacanalis, Perrenoud, 1931, cestoda Lecanicephalida from marine fishes. *Ibid*, 18(11): 133–139.

Shinde, G.B., Mohekar, A.D. and Sarwade, D.V., 1982.Two new species of genus Hexacanalis Perrenoud, 1931. (Cestode : Lecanicephalida from *Trygon zugei*, Muller and Henle land T.Sephani (Cuvier) from Arabian sea west cost of India.

Chapter 24

Length-Weight Relationship in *Mystus seenghala*

☆ *M.S. Pentewar and V.S. Kanwate*

Introduction

Mystus seenghala is a teleost fish belonging to the family Siluridae, one of the most important families of the bony fish. The family siluridae comparises predominantly freshwater fishes that are widely distributed all over the world and it include thirty genera and eighteen species, out of these thirteen genera are found in freshwater through out in India (Day 1971). The length weight relationship has been carried out by many workers like Le Cren (1951), Jhingran (1952), Natarajan and Jhingran (1963), Siddiqui (1977), Hussan and Grover (1986) have investigated the length weight relationships.

The length weight relationship is one of the standard method yielding authentic biological information, by which we can calculated weight from length of fish. The length of fish can be measured more easily and accurately than weight. Determination of such consenting mathematical relationship between length weight of the numerous practical applications in fishey biology. The length weight relationship of most fish can be adequately described by the formula.

$$W = a\,L^b$$

where,

W: Weight,

a: Constant

L: Length and b is exponent.

To study the length weight relationship of *Mustus seenghala* from Sikara dam near Mukhed District Nanded (M.S.) India, hence present investigation was under taken.

Materials and Methods

The Material for present study was obtained from Sikara dam near Mukhed Distrit Nanded (M.S.) India, during January to December 2003. Total 317 Specimens were examined out of which 149 were males and 168 were females. Male *Mystus seenghala* ranged from 9.7 to 17.8 cm in total length and 10.62 to 26.50 gm in weight, where the females ranged between 9.3 to 17.6 cm in total length and 10.23 to 35.62 gm in weight. The males and females are grouped in to 9 size groups with 9mm class intervals. Total length of the fish to the nearest millimeter and carefully noted for this study. The fish was measured from the tip of the closed mouth to the surface water and mucous were removed.

The data for males and females were analysed separately. The general formula $W = a L^b$ has to be fitted in for above data covering this in to logarithms and after substituting Y for log W, X for log L and a for log a the equation will be $Y = a + bX$.

Result and Discussion

The length weight relationship calculated separately for the males and females for different length groups are expressed in Tables 24.1 and 24.2. The regression equations obtained for the two groups are as follows.

$$Y = a + b X$$

Mustus seenghala (*Male*) $Y = -0.4878 + 1.5220 X$

$$W = 0.6883 L^{1.5220}$$

Mustus seenghala (*Female*) $Y = -0.5805 + 1.6246 X$

$$W = 0.7638 L^{1.6246}$$

The value of exponent in *Mustus seenghala* male and female found below 2. In male value of 'b' is 1.5520 and in females 1.6246. From the above equation it is observed that there being slight increase in the value of the exponent in the females than in males as stated by Madlapure (1971) in *Barbus titco*. Similar results were earlier reported by Apparao (1982), Murty (1983) in *Silver belly*, Hoda (1983), Ahirrao (2002) in *M. armatus*.

Table 24.1: Length-Weight Relationship in Male *Mystus seenghala*

Sl.No.	Size Groups (cm)	Average Length in cm L	Average Weight in gm W	Log L X	Log W Y	X^2	XY	Calculated Y	Calculated W
1.	9–9.9	9.7	10.62	0.9868	1.0261	0.9737	1.0125	1.0141	10.33
2.	10–10.9	10.5	11.83	1.0212	1.0730	1.0428	1.0957	1.0665	11.65
3.	11–11.9	11.9	12.55	1.0755	1.0986	1.1567	1.1815	1.1491	14.09
4.	12–12.9	12.35	14.365	1.0916	1.1571	1.1915	1.2571	1.1736	14.41
5.	13–13.9	13.45	16.285	1.1287	1.2116	1.2739	1.3673	1.2301	16.48
6.	14–14.9	14.1875	18.037	1.1516	1.2566	1.3261	1.4464	1.2656	18.41
7.	15–15.9	15.16	22.224	1.1807	1.3468	1.3940	1.5901	1.3092	20.38
8.	16–16.9	16.55	26.145	1.2188	1.4173	1.4854	1.7274	1.3672	23.29
9.	17–17.9	17.5	14.71	1.2430	1.3929	1.5450	1.7313	1.4041	25.36
	Total	121.2975	146.766	10.0979	10.9794	11.3896	12.4099	10.9794	155.4

Table 24.2: Length-Weight Relationship in Female *Mystus seenghala*

Sl.No.	*Size Groups (cm)*	*Average Length in cm L*	*Average Weight in gm W*	*Log L* X	*Log W* Y	*X^2*	*XY*	*Calculated* Y	W
1.	9–9.9	9.3	10.73	0.9685	1.0306	0.9379	0.9981	0.4929	9.837
2.	10–10.9	10.75	12.01	1.0315	1.0795	1.0639	1.1135	1.0953	12.46
3.	11–11.9	11.5	12.93	1.0607	1.1116	1.1250	1.1790	1.1427	13.89
4.	12–12.9	12.5	13.355	1.0969	1.1255	1.2031	1.2345	1.2015	15.91
5.	13–13.9	13.8	17.02	1.1344	1.2309	1.2993	1.4031	1.2714	18.68
6.	14–14.9	14.266	19.453	1.1541	1.2889	1.3319	1.4875	1.2945	19.7
7.	15–15.9	15.175	21.917	1.1810	1.3406	1.3947	1.5832	1.3382	21.79
8.	16–16.9	16.166	27.48	1.2084	1.4309	1.4602	1.7390	1.3821	24.11
9.	17–17.9	17.4	32.15	1.2405	1.5077	1.5388	1.8703	1.4348	27.21
	Total	120.857	167.045	10.0815	11.1544	11.3554	12.6084	11.1538	163.587

References

Ahirrao, S.D., 2002. Study of Mastacembelly armatus freshwater fish. *Ph.D. Thesis*, S.R.T.M.U., Nanded.

Apparao, T., 1982. Length-weight relationship in *Pennahia aeropthalmus* (Bleeker) and *Johnius Carutta* (Bloch). *Indian J. Fish.*, 29(1 and 2): 263–266.

Hoda, S.M.S., 1983. Some observation on the distribution, maturity stages and length-weight relationship of the *Anchovythryssa mystae* in the Northen sea. *Indian J. Fish.*, 30(2): 278–286.

Hussan and Grover, J.H., 1986. Length relationship of newly hatched cultured warm fishes *J. Aqua. Trop.*, 1: 67–74.

Jhingran, V.G., 1952. General length-weight relationship of three major carps of India. *Proc. Nat. Inst. Sci.*, Indian, 8B(5): 449–460.

Le, Cren E.D., 1951. Length-weight relationship and seasonal cycle in gonad weight condition in perch (*Perea fluviatilis*). *J. Anim. Ecol.*, 20: 20–219.

Madlapure, V.R., 1973. Biology of *Barbus Puntius ticto* (Han). *Ph.D. Thesis*. Dr. B.A.M.U., Aurangabad, M.S., India.

Murty, S.V., 1983. Observation on some aspect of biology of silver belly. *Leignathus bindus* (Vol.), Kakinada India. *J. Fish.*, 30(1): 61–68.

Natrajan, A.V. and Jhingran, A.G., 1963. On the biology of *Catla* (Ham) from the river Jamuna. *Proc. Nat. Inst. Sci., India*, 29(3): 326–365.

Siddiqui, A.O., 1977. Reproductive biology length-weight relationship and relative condition of Tilapia, Leyesticta (Trewavas) in lake Naivasha, Kenya. *J. Fish*, 01-10: 25–260.

Chapter 25

Costs and Earnings Analysis of Gill Net and Trawl Net Operation along the Ratnagiri Coast

☆ *S.K. Barve, P.C. Raje, M.M. Shirdhankar, K.J. Chaudhari and M.M. Gawde*

Introduction

Marine fish production is almost stagnant for last three to four years even with increased fishing efforts. Increased fishing efforts have affected the catch per unit effort (CPUE) of fishing units operated along the coast of Maharashtra (Mukharji, 2006) making the business uneconomical. Trawling is high investment business as compare to gill net operation. Many workers have made attempts to estimate the economic viability of the various fishing operations such as trawling (Chidambaran and Soundra, (1991); Sathiadhas *et al.* (1992); Sehara and Kanakkan, (1993); Sehara *et al.* (1994) and gill netting (Rao and Pandey, (1990); Sathiadas and Benjamin, (1990); Markad, (2004) operating in the Indian waters.

Ratnagiri is an important maritime district of Maharashtra along the west coast of India. A total of 3,649 number of fishing crafts are operated along the coast of Ratnagiri district of which 1,785 are mechanised fishing crafts contributing 96,644 mt. In the present study an attempt is made to compare the costs and earnings of trawl and gill net operation along the coast of Ratnagiri with a present scenario of reduced catch per unit efforts.

Materials and Methods

Ratnagiri district, of Maharashtra state with 167 km coastline on the west coast of India was selected for the present study. Dabhol (Lat. 17° 35′ 18″ N and Long. 73° 10′ 36″ E) in the north and Sakharinate (Lat. 16° 37′ 48″ N and Long. 73° 21′ 25″ E) in the south as major landing centers and

Mirkarwada (Lat. 16° 59′ 48″ N and Long. 73° 16′ 48″ E) as minor fishing harbour and Kasarveli (Lat. 17° 02′ 27″ N and Long. 73° 17′ 54″ E) as fish landing center were selected to collect the data for the present study. Data on the costs and earnings of 246 trawlers and 21 of gill-netters were recorded for one year by employing structured interview schedule. Separate interview schedule was formulated for gillnet and trawlers. Interview schedule constituted mainly two sections. First with basic information on one time expenditure towards the construction of vessel and fabrication of net while in second section variable costs and revenue were included. The information in the first section was collected once from respondent under study while the information in the second section was collected weekly from the respondent for the period of one year.

Costs and earnings analysis was carried out for each category separately. Costs incurred for fishing right from construction of boat to revenue generated from sale of catch were averaged separately for both types of fishing operations to estimate the average values of each item. Some of key indicators such as capital-turn-over ratio, gross ratio, variable cost ratio, rate of return on loan amount and pay back period were estimated (Bensan, 1999; Salim and Biradar, 2001).

Result and Discussion

Costs and earning statement of gill netters and trawlers are presented in Table 25.1. Capital investment in the trawling operation was eight times more than that of the gill net operation as the size of vessel and capacity of engine used for trawler were much larger than that of the gill-netters. Total non-recurring investment incurred by gill netters and trawlers was estimated Rs. 2,24,100/– and Rs.16,49,500/– respectively.

Table 25.1: Cost and Earning Statement of Gill Netters and Trawlers

Aggregate Measures	*Gill Netters*	*Trawlers*	*Per cent*
Total capital cost (Rs.)	224100	1649500	86.41
Total variable cost (Rs.)	278822	963292	71.06
Total fixed cost (Rs./yr)	136511	719235	81.02
Total cost (Rs./yr)	415333	1682527	75.31
Total revenue (Gross income)	669505	2140970	68.73
Net profit	254172	458443	44.56
Loan amount	377192	1959594	80.75
Interest	52807	274343	80.75
Depreciation	15000	164950	90.91
	Ratio		
Capital turn over ratio	2.99	1.30	
Gross ratio	0.62	0.79	
Variable cost ratio	0.42	0.45	
Fixed cost ratio	0.20	0.34	
Rate of return to loan amount	81.39	37.39	
Pay back period	0.70	1.84	
Pay back period for loan amount	1.40	3.14	

Variable cost estimated for gill-netters and trawlers was Rs. 2,46,950/– and Rs. 6,35,692/– respectively. Variable costs reported by Sathiadhas and Panikkar, (1989) was Rs. 2,95,964/– for 8.5 m trawler; Rs. 3,70,393/– for 9 m Trawler and Rs. 4,61,458/– for 9.5 to 10 m trawler at Tuticorin while Sehara and Kanakkan, (1993) reported variable cost at Rs. 5,87,461/– for trawlers operating along Kerala coast. Kunjir, (2004) reported operational cost of trawlers as Rs. 12,55,571/-. Rao and Pandey, (1990) estimated variable cost Rs. 1,18,190/– for 21-30 ft gillnetters and Rs. 3,11,425/– for 31-40 ft gillnetters whereas Iyer, (1993) reported variable cost of gillnet operation in range of Rs. 1,24,770 to 1,38,366. Operational cost incurred was estimated Rs. 1,65,470/– by OBM gillnetters and Rs. 3,53,442/– by IBM gillnetters by Kunjir, (2004). The salary of crew, expenditure on grocery (24.67 per cent) and maintenance charges of vessels (4.15 per cent) were the main components to record difference in the variable costs of trawlers and gill-netters. The salary of crewmembers and maintenance charges of boats of trawlers alone are nearly equal to the total variable cost of gill-netters. The money invested for floats and sinkers were sixty per cent more in gill-netters than that of trawlers. Fixed cost calculated for gill-netters was Rs. 1,36,511/– and Rs. 7,19,235/– for trawlers. Fixed cost calculated by Rao and Pandey, (1990) for trawlers was in range of Rs. 52,725/– to Rs. 99,725/– whereas Chhaya *et al.* (1991) reported fixed cost Rs. 1,572/– per trip. Fixed cost estimated by Kunjir, (2004) was in range of Rs. 2,31, 304/– to 3,90,108/– excluding repayment of loan. For 21-30 ft gillnetters Rao and Pandey, (1990) estimated fixed cost as Rs. 23,675/– and Rs. 40,025/– for 31-40 ft gillnetters. Fixed cost incurred by IBM gillnetters reported as Rs. 99,003/– and Rs. 52,746/– for OBM gillnetters studied by Markad, (2004) where he not included repayment of loan in fixed cost. The fixed cost calculated in present study is quite high because the cost of non-recurring expenditure affected the fixed cost in terms of depreciation. Fixed cost estimated for trawlers was 81.03 per cent extra than gill-netters on account of difference in capital investment that leads in 75.34 per cent superfluous total expenditure to trawlers compared with gill-netters. The total production cost for first year was worked out to Rs. 4,15,333/– for 293 days of fishing operation *i.e.* Rs. 1507/– per trip for gill-netters whereas it was Rs. 16,82,527/– for 215 days of fishing operation *i.e.* Rs. 7826/– per trip for trawlers. The total cost per day is 82.38 per cent more for trawlers than that of gill-netters. Chhaya *et al.* (1991) projected total cost of Rs. 6944/– per trip for trawlers operated along Jamnagar district of Gujarat whereas average annual cost of productions reported by Sathiadhas and Benjamin (1990) has ranged between Rs. 2.27/– lakhs to Rs. 5.68/– lakhs in the year 1985–86. While studying costs and earnings of gill net operation, Rao and Pandey, (1990) estimated total cost of Rs.1,14,865 whereas Markad, (2004) reported total expenditure of Rs. 2,18,216/– and Rs. 4,52,445/– for OBM and IBM gill-netters. The total cost estimated in present study is quite high but this is justifiable because of hike in recurring and non-recurring cost over a period of time. Average fishing day along Maharashtra coast for trawlers were 208 to 217 as per the Sehara *et al.* (1991) are comparable with present 215 fishing trip in a year. Net profit gained by gill-netters was estimated Rs. 2,54,172/- while it was Rs. 4,58,443/– for trawlers. The trawlers were getting 68.73 per cent additional revenue than gill-netters but when we consider the net profit gained by the trawlers over gill-netters its only 44.56 per cent despite of 75.34 per cent added total expenditure compared to gill-netters. Chhaya *et al.* (1991) has also claimed the same thing that the cost economics of gill-netter and dugout with out board engines (OBM) are far better than trawler. Chacko (1993) has also concluded that the medium sized boat with out-board engine operating gill nets is more economical than mini trawler and dingy boat with shore seine.

Table 25.2: The Estimated Values of Indicators

Sr. No.		*Particulars of Item*	*Rate in Rupees*	*Quantity*		*Amount in Rupees*	
				Gill-netter	*Trawler*	*Gill-netter*	*Trawler*
I	**Non-recurring cost**						
	1	Cost of construction of boat				100000	1375000
	2	Cost of engine				50000	270000
	3	Cost of otter boards				–	4500
	4	Cost of net				74100	–
		Total				**224100**	**1649500**
II	**Recurring Cost**						
	A	**Variable Cost**					
	1	Cost of net	5750	4			23000
	2	Floats				7500	
	3	Ring				3000	
	4	Net repairs				3000	
	5	Cost of chain and floats					3000
	6	Cost of winch wire bundle					9000
	7	Maintenance					40000
	8	Salary					
		a) Tandel	5000		12		60000
		b) 6 crew members	10800		12		129600
	9	Grocery					48000
	10	Ice @ Rs. 0.50 per kg	400/day		215		43000
	11	Diesel	28/litre	9000 litre	21500 litre	252000	602000
	12	Oil	80	30 litre	25 litre	2400	2000
	13	Grease	50	5 Kg	20 Kg	250	1000
	14	Bottom paint	131	12 kg	12 kg	1572	1572
	15	Tubs	250	12		3000	
	16	Baskets	30		24		720
	17	Kerosene	10	600 litre		6000	
	18	Licence Fee				50	250
	19	Port Fee				50	150
		Total				**278822**	**963292**
III	**Total Project cost**					**502922**	**2612792**
IV	**Total Loan Amount**					**377192**	**1959594**

Contd...

Table 25.2–Contd...

Sr. No.		Particulars of Item	Rate in Rupees	Quantity		Amount in Rupees	
				Gill-netter	Trawler	Gill-netter	Trawler
II	B	**Fixed Cost**					
	1	14% interest on loan amount				52807	274343
	2	Repayment–1/7 of loan amount				53885	279942
	3	Depreciation					
		a) 10% depreciation on boat and engine				15000	164950
		b) 20% depreciation on net				14820	
		Total				**136511**	**719235**
V		**Total expenditure**				**415333**	**1682527**
VI		**Revenue (293 days for gill-netters and 215 days for trawlers)**		**2285**	**9958**	**669505**	**2140970**
VII		**Profit**				**254172**	**458443**

Economic indicators were calculated from costs and earning data. The estimated values of indicators are given in Table 25.2. This indicates that gill-netters were getting gross revenue of Rs. 2.99/– for each rupee of capital investment whereas trawlers were receiving only Rs. 1.30/– for each rupee of capital investment even though 86.41 per cent more initial investment against gill-netters. The capital turn over ratio has ranged from 2.68 to 4.73 of 32 ft trawler (Sathiadhas and Benjamin, 1990), 2.56 for 9.5 m trawler operated along the Nagapattinam coast (Sathiadhas *et al.*, 1992). For gill net operation, Sathiadhas and Benjamin, (1990) found capital turn over ration of 2.58 at Cuddalor and 1.80 at Nagapattinam whereas Sathiadhas *et al.*, (1991) estimated capital turn over ratio 1.61 for gill-netters operated along Tuticorin coast. This reports suggest that capital turn over ratio is decrease with time for trawlers because increase in capital investment and almost stagnant catch per unit effort. The gross ratio suggests that gill-netters were spending 60 paise while trawlers were spending 79 paise out of one rupee earned towards total expenditure. The variable cost ratio for gill-netters and trawlers advocate that both were spending nearly equal amount of their net income towards operational cost but difference in fixed cost ratio making gill net operation more profitable considering initial investment. Rate of returns on loan amount indicates that gill-netters could recover almost their entire loan amount within first year itself but for trawlers it would take three years. Payback period for initial investment was worked out as 0.67 years for gill-netters and 1.84 years for trawlers. Payback period for loan amount estimated about one and half years for gill-netters while it was slightly more than three years for trawlers. The economic indicators projected by various investigators like Sathiadhas and Pannikar, (1989); Sehara and Karbhari, (1989a); Sathiadhas and Benjamin, (1990); Sathiadhas *et al.* (1991); Sehara and Kanakkan, (1992); Senthilathiban *et al.* (1997) can be compare with present results.

Thus from the present research we can say that trawlers are getting more profits but compared with the returns on unit investments gill-netters are far excellent than that of trawlers.

Chapter 26

Study of Protein Metabolism in Hepatopancreas and Muscle of Prawn *Penaeus monodon* on Exposure to Altered pH Media

☆ *V. Sailaja, E. Madhuri, K. Ramesh Babu, S. Rama Krishna and M. Bhaskar*

Introduction

Proteins are the most important group of macromolecular substances which occupy a pivotal place in both structural and dynamic aspects of living matter (Harper *et al.*, 1983). Proteins contribute to the process of interactions between intra and extra cellular media. The variations in the experimental conditions inturn influence the protein enzyme interactions within the cellular organization. Proteins play a significant role in the permeability properties of membranes which control the osmotic and ionic regulatory processes (Florkin *et al.*, 1964: Venkata Reddy, 1976). There is considerable evidence implicating the importance of free amino acids in intracellular osmoregulation in invertebrates (Florkin and Schoffeniels, 1969; Bedford, 1971; Kasschau, 1975) and in several teleostean fishes (Lasserre and Gilles, 1971; Colley *et al.*, 1974; Ahokas and Duerr, 1975; Ahokas and Sorg, 1977; Jurss, 1980).

Variations due to alterations in experimental conditions lead to a variety, of changes in different protein fractions and these variations in turn influence the protein enzyme interactions within the cellular organization (Brattstrom and Lawrence, 1962). Higher protein degradations through increased protease activity of tissues were reported in various physiological and pathological conditions (Narayana Reddy, 1976). Bhaskar *et al.* (1983) reported that the hepatic tissue proteolysis of freshwater fish on exposure to altered environmental pH while Sobha Rani *et al.* (1983) reported the effect of altered pH on white muscle protein fractions in *Tilapia mossambica*.

All these findings once again emphasize the participation of proteins in the alteration of the internal environment to counteract changes in the external environment. But none of the studies have considered the effect of environmental acidification on tissue protein metabolism which deals with both synthesis and degradation of proteins during the exposure to altered pH in prawn. Hence, it is felt desirable to under take tissue protein metabolism in the present study in order to understand the changes in the protein metabolism of hepatic and muscle tissues of prawn in response to imposed acidic and alkaline media.

Materials and Methods

The tiger prawns, *Penaeus monodon* were obtained from magunta prawn hatchery at Nellore, A.P., India and they were maintained under laboratory conditions in normal brackish water at room temperature (27.5°C±0.5°C), Salinity (25 ppm), pH (7.4±0.1) and exposed to 12 hrs dark and 12 hours photoperiod. The prawns were fed daily with a standard commercial diet. The prawns with the standard size and weight such as 6.0±0.5 gm were selected for the study. After acclimatization to the laboratory condition the prawns were divided into 3 batches and each batch consists of 10 prawns. The first batch considered as control and maintained them in normal brackish water pH 7.4±0.1. The second and third batch of prawns were exposed to acidic media at pH 6.5±0.1 and alkaline media at pH 9.0±0.1 for 24 hours respectively and treated them as acute exposure (experimental).

A special standard device described by Bhaskar and Govindappa in 1986 to maintain the pH of the acidic and the alkaline media has been used in the present study. After 24 hours of acute exposure, both hepatopancreas and muscle tissues of control and experimental prawns were isolated, chilled and used for tissue biochemical analysis. The free amino acid content was estimated by the method of Moore and Stein (1954) as described by Colowick and Kaplan (1957). The protein content was estimated by the method of Lowry *et al.* (1951). The enzymes like ACP and ALP were estimated by the method of Bodansky (1933). The AAT and AlAT enzyme activity were estimated by the method of Reitman and Frankel (1957) as described by Bergmeyer (1965).

Results

The changes in protein fractions were observed in the hepatopancreas of prawn on exposure to acidic and alkaline media (Table 26.1). Soluble and structural proteins were decreased significantly in both acidic and alkaline media when compared to control. The free amino acid levels were elevated significantly in the hepatopancreas of prawn on exposure to acidic medium over control. In contrast, the same was depleted in alkaline medium than control. The enzyme activity levels of AAT, AlAT, ACP and ALP were significantly increased in the hepatopancreas of prawn on exposure to both acidic and alkaline media over control. The soluble proteins/total proteins (SP/TP), the structural proteins/ total proteins (StP/TP) ratios were lowered in both acidic and alkaline media.

Soluble and structural proteins were significantly decreased in hepatopancreas of prawn on exposure to acidic medium. However, the free amino acid levels were significantly increased over control. The pattern of enzymes such as AAT, AlAT, ACP and ALP were prominently elevated in the hepatopancreas of prawn on exposure to acidic medium. However, the (SP/TP) and (StP/TP) ratios were lowered when compared to control.

Table 26.1: Changes in Protein Fractions and Enzyme Activity Pattern in Hepatopancreas on Acute Exposure to Altered pH Media

Sl.No.	*Component*	*Control*	*Acidic (pH 6.5)*	*Alkaline (pH 9.0)*
1.	Soluble Proteins (mg/gm wet wt. of tissue)	50.80±4.26	44.32±4.07 –12.75 P<0.001	42.77±3.49 –15.81 P<0.001
2.	Structural Proteins (mg/gm wet wt. of tissue)	42.85±3.53	36.38±3.11 –15.10 P<0.001	37.85±2.79 –11.67 P<0.001
3.	Free Fatty Acids (µ moles/gm wet wt. of tissue)	20.30±1.89	29.37±2.45 +44.68 P<0.001	18.90±1.32 –6.90 P<0.01
4.	AAT (µ moles of sodium pyruvate formed/mg protein/hr)	1.51±0.096	2.05±0.162 +35.76 P<0.001	2.55±0.178 +68.87 P<0.001
5.	AlAT (µ moles of sodium pyruvate formed/mg protein/hr)	2.21±0.163	2.78±0.213 +25.79 P<0.001	3.14±0.247 +42.08 P<0.001
6.	ACP (µ Pi liberated/mg protein/hr)	2.98±0.153	3.72±0.315 +24.83 P<0.001	4.08±0.340 +36.91 P<0.001
7.	ALP (µ Pi liberated/mg protein/hr)	11.06±0.891	13.87±1.16 +25.41 P<0.001	14.73±1.02 +33.18 P<0.001
8.	Soluble Proteins/ Total Proteins (S.P./T.P.)	0.583±0.036	0.553±0.031 –5.14 NS	0.389±0.026 –33.28 <0.001
9.	Structural Proteins/ Total Proteins (StP./T.P.)	0.492±0.023	0.454±0.022 –7.72 P<0.001	0.354±0.028 –29.88 P<0.001

Each mean value is an average of 6 individual observations. Mean,± S.D. + or – indicate the per cent increase and decrease over control and 'P' denotes the level of statistical significance.

The levels of soluble structural and free amino acids were decreased in the hepatopancreas of prawn on exposure to alkaline medium than control. The activity levels of the enzymes such as AAT, AlAT, ACP and ALP were significantly elevated in hepatopancreas of prawn on exposure to alkaline medium over control. The (SP/TP) and (StP/TP) ratios were also significantly lowered in alkaline medium in comparison to control.

Soluble and structural protein levels of hepatopancreas of prawn were significantly decreased on exposure to both acidic and alkaline media. However the free amino acid levels showed differential pattern. They were increased on exposure to acidic medium and decreased in alkaline medium than control. The activity levels of enzymes such as AAT, AlAT, ACP and ALP were significantly elevated in the hepatopancreas of prawn on exposure to altered pH media over control. The rate of increase was more in alkaline medium than in acidic medium. The soluble/total protein ratio was showed no prominent change in acidic medium, with significant decrease in alkaline medium. The structural/ total protein ratio was also lowered significantly in both acidic and alkaline medium than control. However the rate of reduction was more in alkaline medium than in acidic medium.

Table 26.2: Changes in protein fractions and enzyme activity pattern in muscle tissue on acute exposure to altered pH media

Sl.No.	Component	Control	Acidic (pH 6.5)	Alkaline (pH 9.0)
1.	Soluble Proteins (mg/gm wet wt. of tissue)	42.85±3.16	33.23±3.89 −22.45 P<0.001	38.17±3.06 −10.92 P<0.001
2.	Structural Proteins (mg/gm wet wt. of tissue)	64.34±5.17	42.60±3.49 −33.79 P<0.001	52.90±4.55 −17.78 P<0.001
3.	Free Fatty Acids (μ moles/gm wet wt. of tissue)	19.50±0.0986	24.98±1.66 +28.10 P<0.001	16.72±1.45 −14.25 P<0.001
4.	AAT (μ moles of sodium pyruvate formed/mg protein/hr)	1.86±0.132	2.29±0.184 +23.12 P<0.001	2.68±0.213 +44.09 P<0.001
5.	AlAT (μ moles of sodium pyruvate formed/mg protein/hr)	2.53±0.120	3.32±0.250 +31.22 P<0.001	2.97±0.244 +17.39 P<0.001
6.	ACP (μ Pi liberated/mg protein/hr)	3.92±0.261	4.68±0.340 +19.38 P<0.001	5.03±0.419 +28.32 P<0.001
7.	ALP (μ Pi liberated/mg protein/hr)	12.37±1.05	14.56±1.23 +17.70 P<0.001	15.96±1.38 +29.02 P<0.001
8.	Soluble Proteins/ Total Proteins (S.P./T.P.)	0.369±0.027	0.407±0.031 +10.30 P<0.001	0.307±0.028 −16.80 P<0.001
9.	Structural Proteins/ Total Proteins (St.P./T.P.)	0.554±0.036	0.521±0.039 −5.96 P<0.01	0.425±0.033 −23.28 P<0.001

Each mean value is an average of 6 individual observations. Mean ± S.D. + or – indicate the per cent increase and decrease over control and 'P' denotes the level of statistical significance.

The proteins fractions of prawn muscle were observed on exposure to altered pH media (Table 26.2). The soluble and structural proteins were decreased significantly in muscle of prawn on exposure to both acidic and alkaline media over control. In contrast, the free amino acid content in prawn muscle was increased significantly on exposure to acidic medium but decreased in alkaline medium when compared to control.

Discussion

The soluble and structural proteins were significantly decreased in hepatopancreas of prawn on exposure to both acidic and alkaline media. However, the percentage of depletion was more in soluble protein fraction on exposure to alkaline medium than in acidic. These observations suggest that, hepatopancreas has showed more mobilization of soluble protein fractions than structural proteins towards the catabolic metabolism on exposure to alkaline medium. The depletion in soluble and structural protein fractions suggesting the possibility of tissue proteolysis and/or inhibited protein biosynthesis. The acid proteases were significantly elevated on exposure to acidic medium in liver of fish which might be due to an excess influx of protons into the acidic medium (Bhaskar *et al.*, 1983; Bhaskar, 1994) and the protons activate, the proteases (Bhaskara Haranath, 1979; Murthy, 1981), hence the observed degradation in protein levels might also be due to the active proteolytic activities. The free amino acid levels showed differential pattern on exposure to both acidic and alkaline media. The amino acid content was significantly elevated which supports the activation of proteases as reported by earlier investigators in fish (Bhaskar *et al.*, 1983; Bhaskar, 1994). Another possibility for such an elevation in hepatic tissue amino acid pool might be due to active uptake from the blood. Inspite of protein breakdown of the tissue the free amino acid levels were significantly depleted on exposure to alkaline medium suggesting their active mobilization into other metabolic activities. In view of increased activity of AAT and AlAT in both acidic and alkaline medium, the active involvement of amino acids in transmination and oxidative reactions can be expected. The percentage of increase in the activity levels of AAT and AlAT was much higher in alkaline medium than in acidic medium. Similarly, the acid phosphatase and alkaline phosphatase also showed significant elevation in both acidic and alkaline medium which suggest the active formation of sugar phosphates required for the energy release necessary to counteract the imposed pH stress. Since acid phosphatase is involved in the hexose phosphorylation reaction, the alkaline phosphatase is involved in the transphosphorylation reactions, the elevated ACP and ALP activities might be due to the increased coupling of oxidative phosphorylation reactions leading to the formation of hexose phosphates or accelerated activity of lysosomes and/or their increased formation in response to induced pH stress. The percentage increase of the enzyme activities were much greater in alkaline medium than in acidic medium. The SP/TP ratio showed no significant change in acidic medium, where as in alkaline medium it was lowered suggesting the mobilization of more soluble proteins than total proteins. The StP/TP ratio was also lowered in both acidic and alkaline media but the percentage of reduction was much greater in alkaline medium than in acidic medium.

The muscle protein metabolism in response to acute exposure to acidic and alkaline stress exhibited differential pattern. The soluble and structural proteins were decreased on exposure to both acidic and alkaline media, however the percentage of soluble protein depletion was more in acidic medium than in alkaline medium. Similarly, the structural proteins also decreased drastically in acidic medium when compared to alkaline medium. Decrease in structural protein contents suggest that the tissue protein reserves were being mobilized for other metabolic activities. Since the altered pH media imposes heavy stress on animals (Sobha Rani *et al.*, 1999; Murthy *et al.*, 1999) which requires more energy to counteract the imposed stress, mobilization of these tissue proteins towards the energy production can be expected. The free amino acid content was elevated on exposure to acidic medium, with a decrease in alkaline medium. The elevated level of free amino acid content supports the degradation of soluble and structural proteins in acidic medium. Another possibility for such an elevation in muscle tissue might be due to elevated uptake of amino acids from the blood. The elevated plasma amino acids under stress conditions (Paul *et al.*, 1978) observed by earlier workers supports the uptake of amino acids from blood. Inspite of the depleted protein content the free amino acid content was reduced in alkaline medium which suggest the mobilization of amino acids into other metabolic activities and/or their efflux into the blood. In view of increased AAT, and AlAT activities in both acidic and alkaline media, the active involvement of amino acids in transmination and oxidative reactions can be expected. The percentage increase of AAT activity was greater in alkaline medium than in acidic medium, where as the AlAT activity showed differential pattern and the percentage of increase was much more in acidic medium than in alkaline medium. Since, the AlAT activity is a marker for the mobilization of Alanine, one of the glycogenic amino acid, towards the production of carbohydrates and the increased AlAT in the tissue suggests the stepped up glycogenic activities.

Such a type of tissue metabolism might have been responsible for the increased glycogen content observed in prawn muscle tissue on exposure to both acidic and alkaline media. (unpublished data). These results were in consonance with the earlier reports observed in fish muscle (Murthy, 1981). The ACP and ALP activities were also elevated on acute exposure to both acidic and alkaline media which suggest the active formation of hexose phosphates under imposed pH stress. The percentage of increased of ACP and ALP activities were greater in alkaline medium than in to acidic medium. The SP/TP ratio showed differential pattern on exposure to both acidic and alkaline media. The ratio was higher in acidic medium and lower in alkaline medium which envisages the mobilization of soluble proteins was less in acidic medium than in alkaline medium. The StP/TP ratio was lowered in both acidic and alkaline media, but the rate of depletion was greater in alkaline medium than in acidic medium.

Finally it is concluded that the acute exposure to altered pH media induced significant alterations in protein metabolism leading improved survival through metabolic adjustment.

References

Ahokas, R.A. and Duerr, 1975. Tissue water and intracellular osmoregulation in two species of euryhaline teleosts, *Culaea inconstans* and *Fundulus diaphanous Comp. Biochem. Physiol.*, 52A: 449–454.

Ahokas, R.A. and Sorg, G., 1977. The effect of salinity and temperature, on intracellular osmoregulation and muscle free amino acids in *Fundulus diczphanus. Comp. Biochem. Physiol.*, 56A: 101–105.

Bedford, J.J., 1971. Osmoregulation in Melanopsis trifaciata Gray 1843–III. The intracellular nitrogenous compounds carp. *Biochem. Physiol.*, 40A: 899–910.

Bergmeyer, H.U., 1965. In: *Methods of Enzymatic Analysis.* Academic Press, New York and London.

Bhaskar, M., 1994. Changes in the liver protein fractions of *Tilapia mosambica* (Peters) during acclimation to low and high pH media. *Fisheries Research*, 19: 179–186.

Bhaskar, M., Manohar Reddy, R., Chalapathy Rao, M.V., Sobha Rani, P. and Govindappa, S., 1983. Pattern of hepatic tissue proteolysis of freshwater fish on exposure to altered Environmental pH. *Indian J. Comp. Anim. Physiol.*, 1(2): 47–52.

Bhaskara Haranath, V., 1979. Studies on some aspects of starvation induced metabolic modulations in selected tissues of *Tilapia mossambica* (Peters). *Ph.D. Thesis*, S.V. University, Tirupati, A.P., India.

Bodansky, A.I., 1933. Determination of inorganic Phosphate. Beer's Law and interfacing substances in the 'Kuttner–Lichtenstein Method. *Biol. Chem.*, 99: 197–206.

Brattstrom, B.H. and Laurence, P., 1962. The rate of thermal acclimation in Anuran amphibians. *Physiol. Zool.*, 35: 148–156.

Colley, L., Fox, F.R. and Huggins, A.K., 1974. The effects of change in external salinity on the non-protein nitrogenous constituents of parietal muscle from *Agonus cataphractms. Comp. Biochem. Physiol.*, 48(A): 757–763.

Colowick, S.P. and Kaplan, N.O., 1957. In: *Methods in Enzymology*. Academic Press, New York, 3: 501.

Florkin, M., Duchatean-Bossen, G.H., Jeuniau, C.H. and Schofi'e-Niels, E., 1964. Sur le mechanlsme dela regulation de la concentration intracenulaire on acids, amines, libres chez friocheir sinensis, an tours de l adaptation osmotique *Arch. Internat. Physiol. Biochem.*, 72: 802–906.

Florkin, M. and Schoffeniels, F., 1969. *Molecular Approaches to Ecology*. Academic Press, New York.

Harper, A.H., Rodwell, W.V. and Mayes, A.P., 1983. *Review of Physiological Chemistry*, 19th edn. Lange Medical Publications. Maruzen Asia (Pvt.) Ltd., Singapore, p. 647–655.

Jurss, K., 1980. The effect of changes in external salinity on the free amino acids two amino transferases of white muscle from fasted Salmo gairdneri (Ttiehardson) *Comp. Biochem. Physiol.*, 65(A): 501–504.

Kasschau, F.P., 1975. The relationship of free aminoacids to salinity changes and temperature-salinity interactions in the mud-flat snail *Nassarious obsolutus. Comp. Biochem. Physiol.*, 20(A): 301–308.

Lasserre, P. and Gilles, R., 1971. Modification in the amino acid pool in the pariental muscle of two euryhaline teleosts during osmotic adjustment. *Experentia*, 27: 1434–1435.

Lowry, O.H., Rosebrough, N.J., Farr, A.L. and Randall, R.J., 1951. Protein measurement with the Folin Phenol reagent. *J. Biol. Chem.*, 193: 265–275.

Moore, S. and Stein, W.H., 1954. A modified ninhydrin reagent for the photometric determination of amino acids and related compounds. *J. Biol. Chem.*, 211: 907–913.

Murthy, V.K., 1981. Studies on the tissue metabolic modulations in *Tilapia mossambica* (Peters) during the course of acclimation to sub-lethal acidic water. *Ph.D. Thesis*, S.V. University, Tirupati, A.P., India.

Murthy, V.K., Madhuri, E. Sailaja, V., Bhaskar, M. and Govindappa, S., 1999. Studies on the red muscle lipid metabolism of the fish, *Sarotherodan mossamblcus* (Trewavs) on exposure to sub-lethal acidic media. *J. Aqua. Biol.*, 14(1 and 2): 83–85.

Narayana Reddy, K., 1976. Some aspects of denervation muscle atrophy with particular reference to protein metabolism in the frog, *Rana hexadactyla* (Lesson). *Ph.D. Thesis*, S.V. University, Tirupati.

Paul, H.S., Harbhajan, S. and Adibi, S.A., 1978. Leucine oxidation in diabetes and starvation effect of Ketone bodies on branched-chain aminoacid oxidation *in vitro. Meta. Clin. Exp.*, 27(2): 185–200.

Reitman, S. and Fraenkel, S., 1957. A calorimetric method for the determination of glutamic, oxaloacetic and glutamic pyruvic transaminases *Amn. Clin. Pathol.*, 28: 56–58.

Sobha Rani, P., Bhaskar, M, Krishna Murthy, V., Reddanna, P. and Govindappa, S., 1983. Effect of environmental acidity and alkalinity on white muscle protein fractions of freshwater fish, *Tilapia rnossambica* (Peters). *J. Environ. Biol.*, 4(3): 155–160.

Sobha Rani, P., Madhuri, E., Sailaja, V., Bhaskar, M. and Govindappa, S., 1999. Changes in the excretory products of fish red muscle on acclimation to low pH. *J. Aqua. Biol.*, 14(1 and 2): 87–89.

Venkata Reddy, V., 1976. Studies on mechanisms underlying acclimation to salinity in the freshwater crab *Paratelphysa hydrodromus* (Herbst). *Ph.D. Thesis*, S.V. University, Tirupati.

Chapter 27

Reproductive Biology of *Xancus pyrum* Varieties Acuta and Obtusa from Thondi Coastal Waters-Palk Strait (South East Coasts of India)

☆ *C. Stella and J. Siva*

ABSTRACT

The chank *Xancus pyrum* varieties are marine in distribution. Chank prefer fine and soft sandy areas called locally 'poochi–manal' or pirals. They are distributed in the seafloor in depth of 10 to 20 meters in sand mixed with mud. The sacred chank form are very important. Commercial fishery in the Gulf of Mannar and Palk Strait. During breeding season when the capsules are fashioned and rooted in the sand extends throughout June to March. From the available literature, it is known that no intensive work has been done on the varieties of *Xancus pyrum*. An attempt has been made to study breeding biology of two varieties of *Xancus pyrum* (var) *acuta* and *Xancus pyrum* (var) *obtusa*.

Introduction

The sacred chank forms are very important commercial fishery in the Gulf of Mannar and Palk Strait. Hornell (1914, 1916 and 1922) has described the varieties and races of the chanks occurring in Indian and Ceylon waters. During breeding season when the capsules are fashioned and rooted in the sand extends throughout January to March. In the recent years our knowledge of the eggs and larvae of marine gastropods has increased considerably.

The breeding biology include egg masses and larval developments of some prosobranchs have been studied by Risbec (1928, 1931, 1932, 1935). Hornell (1921) has described eight egg masses and

Gravely (1942) described twelve egg masses. A few details are also available on the egg masses of *Thais carnifera* (Annandale and Kemp, 1916) *Potamides cingulates* and *Natica* (Panikkar and Aiyer, 1939) Tonna sp. (Panikkar and Tampi, 1949) and *Thais bufo* (Chari, 1950) and on larval development of *Xancus pyrum* (Chidambaram and Unny, 1947). From the above information it is clear that the studies on the breeding habits and the larval development of marine prosobranchs have not received much attention from the Indian workers. Krusadai, Pamban and Mandapam are well known for their rich molluscan fauna and it was suggested that the breeding biology of them will be of considerable scientific interest. Hence an attempt was made to study on the breeding biology of the two species namely, *Xancus pyrum* var. *acuta* and *Xancus pyrum* var. *obtusa.*

Materials and Methods

The specimens were collected from the Thondi coast, Palk Strait. The spawn masses are usually found on the under surfaces of the stones, boulders or upon the algae, dead shells, corals and crevices of the rocks. Some of these spawn masses themselves serve for the attachment of the spawn of the others. The adult animals were kept in aquaria to observe the spawning habits. Breeding season was determined as far as possible for each species under investigation.

The specimens were collected monthly and their reproductive condition was recorded based on the colour of the gonad and the extent to which it is spread over the digestive gland. In two species, the sex ratio was studied by random collection of specimens. The size at which sexual maturity was attained in males and females was determined by observing the maturity of accessory sexual organs and by the presence of matured eggs and sperms in the smears of the gonads. Copulation was observed in the laboratory. The copulated pairs were selected for observations on spawning and developmental studies. Shell height was measured with the aid of a vernier caliper to the nearest 0.01mm under a compound microscope.

The copulation and egg laying behaviour were observed for two species through the aquarium wall. Copulated females were kept in glass troughs filled with sea water of the study area and spawning was observed at the room temperature (29±2° C). The salinity of water in the glass trough was maintained at 30±2 ppt in order to provide the proper environmental conditions for the developing embryo. The water was renewed daily.

The fecundity of each species which was kept in different aquarium tanks was determined by counting the number of egg capsules laid in the aquarium tank. The morphology of the individual capsules was studied under a dissection binocular microscope and the number of eggs in each capsule was counted. The number of eggs in each capsule was counted for each species. In a freshly laid capsule, the development of egg was observed.

Results and Discussion

Copulation

In both *Xancus pyrum* var. acuta and *Xancus pyrum* var. *obtusa* copulation was confined from December to July months. In the laboratory copulation was observed both during day and night times. Copulation proceeded spawning at the time intervals of one day on occasions, females were observed to spawn immediately after copulation. In two species the number of copulating pairs showed an increase in early summer and only a few copulatory pairs were observed during pre monsoon months.

Spawning

Under natural conditions the processes of gametogenesis is completed before the normal time for spawning. Spawning in two species is confined to the new and full moon as there was a sudden increase in the number of egg capsule during these periods. Number of eggs per capsule varies from 250 to 350 in *Xancus pyrum* var. *acuta* so also in *Xancus pyrum* var. *obtusa.*

Fecundity

In two species, the capsule were laid in masses. Both in *Xancus pyrum* var. *acuta* and *Xancus pyrum* var. *obtuse,* each capsule contained more than 250 to 350 eggs. There was great difference in the number of egg capsule produced by a captive female depending upon the size, feeding and environmental factors.

Formation of Egg Capsule (Figure 27.1)

In both the species, the egg case is an elongated, loosely, spirally twisted structure and is aptly compared to a miniature ram's horn. It has a base by which it is attached to the sandy substratum. Each egg case is a compound capsule and it is made up of distinct chambers or capsules. The chamber has a dome shaped upper part with a thin walled floor which is surrounded by a broad one above the other. They are attached to a membrane on one side and free on the opposite side. The floor of one capsule fits in compactly on the dome of the lower capsule. In the floor of each chamber is present a cresentic or nearly straight slit. The slit is probably the preformed exit passage. It is covered by a thin transparent membrane. The chambers are not of uniform size. The chambers at the proximal end are small and increase gradually in size upwards till about ¾ of the egg case and afterwards there is a slight decrease in size upto the end. Usually 7-8 chambers from below and 2-4 from above do not hold any embryos. The breadth of the chamber varies between 1.3 and 2.8 cm.

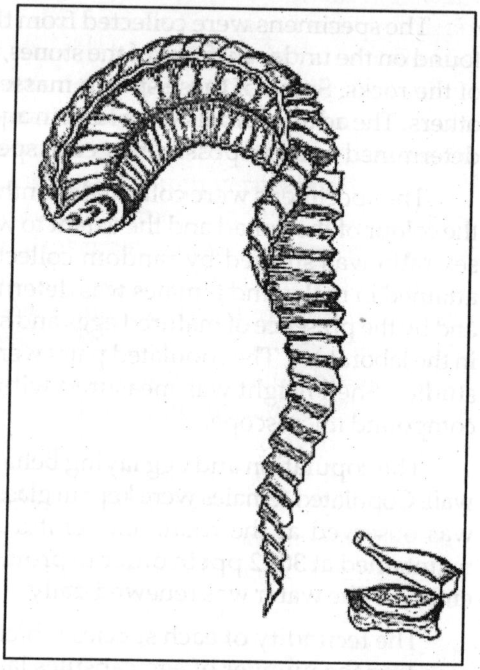

Figure 27.1: Diagram Showing the Egg Capsule of Chank-Xancus Pyrum

At the time of examination the young ones showed about 4 whorls. The first 3½ whorls form the larval shell and the rest beginning to show the adult features. The first 3½ whorls are smooth, plain and colorless. The last whorl begins to show few pinkish brown spots, with very feeble spiral striations. The shell siphon is distinct and the aperture is elongate. In the withdrawn condition with the foot slightly pushed out. The operculum is brownish in colour. The shell measures about 9 mm. in length and 3 mm. in breadth. The young ones thrived at the laboratory for a day.

According to Hornell, the egg capsules are found from January to April, but rarely in May. In the present observation Ram's horns were found in the Thondi coast in the month of July. The egg capsule consists of 25 to 30 or more chamber and about 200 to 250 baby chanks come out of one capsule. The first two upper chambers and the last 6 or 7 chambers did not contain any. The other chambers contained 3 or 6 in each. Each baby chank which escape out of the capsule consists of 3 or 4 whorls of transparent and delicate protoconch, and one whorl of the normal shell. The length of each was 9 mm

Figure 27.2: Diagram Showing Stage I of Veliger Larvae of Chank-Xancus Pyrum

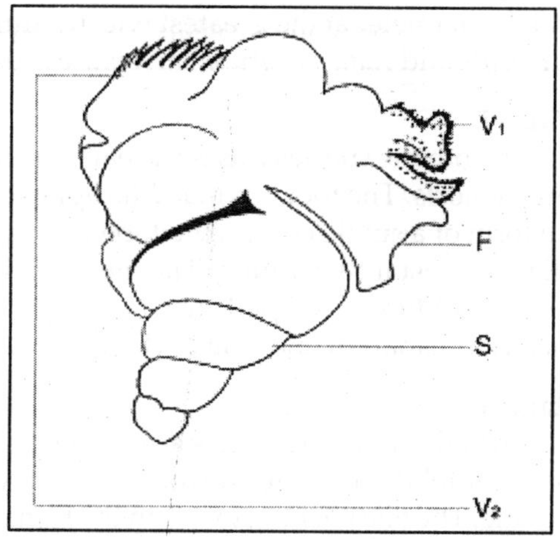

Figure 27.3: Diagram Showing Stage II of Veliger Larvae of Chank-Xancus Pyrum

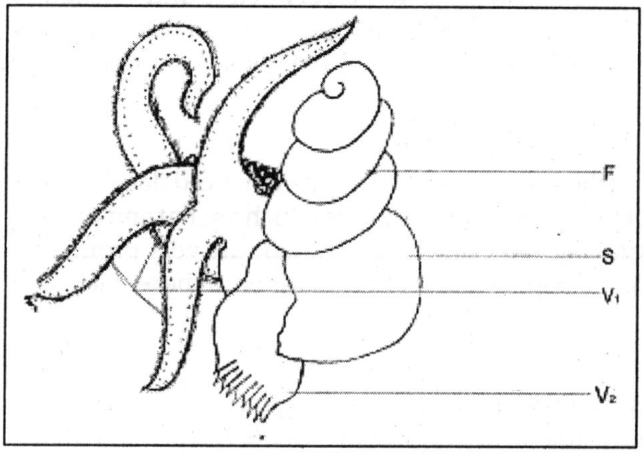

Figure 27.4: Diagram Showing Stage III of Veliger Larvae of Chank-Xancus Pyrum

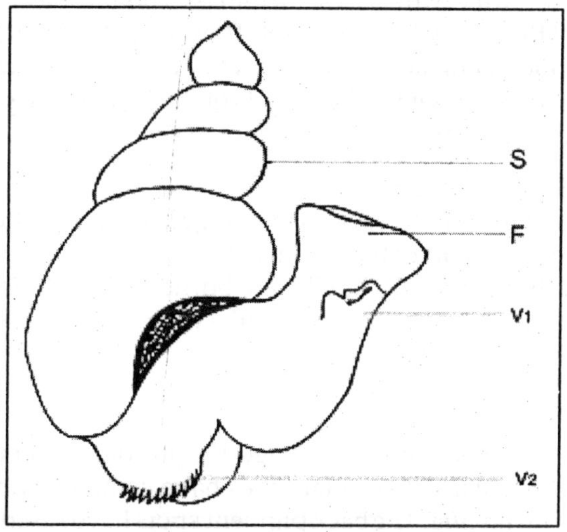

Figure 27.6: Diagram Showing Stage V of Veliger Larvae of Chank-Xancus Pyrum

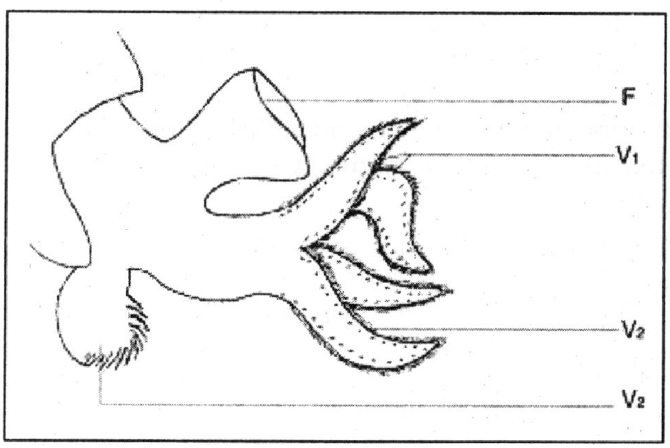

Figure 27.5: Diagram Showing Stage IV of Veliger Larvae of Chank-Xancus Pyrum

List of abbreviations used in the figure and their expansion

F: Foot; S: Shell;
V_1: First part of the velum near the foot
V_2: The other part of the velum away from the foot

and the diameter at the greatest width 5 mm. In the present study, the veliger larvae of *Xancus pyrum* var. *acuta* and *Xancus pyrum* var. *obtuse* is described in detailed manner (Figures 27.2–27.6).

Stage–I

In both the species this is the earliest stage obtained. The larva has an embryonic shell formed of three whorls. The foot has made its appearance. The velum is in two parts; the one near the foot is in the form of a single contractile lobe with a slight depression in the middle and carrying cilia. These cilia are constantly moving. The next part is separated from the first by a contractile lobe which is in the form of finger–shaped structures without cilia. These are also contractile. The first whorl of the shell is transparent and fragile (Figure 27.2).

Stage–II

In both the species this stage is 24 hours older than the previous stage. In this stage the foot and the velar lobes are better developed. The velum possess cilia which are very active and constantly moving. The one near the foot is in the form of three elongated lobes with cilia. A row of black pigment spots has made its appearance on either margin of each lobe. The first three whorls of the shell are transparent and fragile (Figure 27.3).

Stage–III

In the two species stage-III is 24 hours older than the second stage. A very pronounced development of the velar lobes is evident. They are in the form of four prominent tentacles which are flat and possess much elongated structures. They are always contracting and expanding. The cilia are very prominent and are present on both parts of the velum. The pigment spots found at the margins of the tentacles are very well pronounced (Figure 27.4).

Stage–IV

In both the species this stage is obtained after the third day of the first stage. The well developed foot, the part of the velum in the form of four tentacle like structure and the less developed part of the velum, which lacks cilia are borne in the interior part of the body. It is often thrust out in the living condition. But the tentacle–like velar lobes have become shorter in length and less emphasized than in the previous stage (Figure 27.5).

Stage–V

In both the species stage-V is 48 hours older than the previous stage. This is the last stage obtained. The tentacle–like velar lobes have completely disappeared, leaving behind a dark coloured fringe with no cilia. The back pigment spots have fused together to form an irregular patch of the fringe. The foot is very well developed (Figure 27.6).

The trochosphere stage in the development of the chank therefore assumes a modified veliger form characterized by a spiral embryonic shell and a well developed velum. The main part of which is in the form of four long contractile lobes. The creeping foot is fully and normally developed and even after that, the velum persists, as found in the case of veliger larvae known as "Macgillivrayia" and "Chelotropis" of Dolium and purpura respectively. The veliger larval period of *Xancus pyrum* var. acuta and *Xancus pyrum* var. obtuse extends to about five days.

The egg masses of marine prosobranches are very characteristic in their shape and appearance. The following types have been studied in the previous investigation such as gelatinous ribbons (*Certhidea fluviatilis, Cerithium morus*), peculiar sand encrusted collar-like structures (*Natica marochiensis*), large slimy wavy bands (*Tonna dolium*) and capsules of various shapes and sizes (*Murex virgineus* var. *Ponderosa*, *Thais bufo*, *Nassa jacksoniana*, *Ancilla* sp., *Conus araneosus* etc).

Embryos with non-pelagic development may feed in different ways. In all species with these types of feeding the embryonic development seems to be uniform in that the embryos leave the egg capsules after reaching a certain size and organization. Thus all the hatching young of the same species will be at the same stage of development. However this is not the case when the embryos resort to nurse egg feeding. Eggs of more or less uniform size are laid together with in a common egg space or capsule. Only some of the eggs develop into embryos while the rest serve as nurse eggs for the developing ones. The latter will not hatch out until they have exhausted all the nurse eggs.

In the present study, in two species, the nurse egg feeding is worth mentioning. The embryos at early stage were dissected out from the egg capsules and kept in a watch glass along with nurse eggs. They reach the eggs and each egg is manipulated towards the mouth with the help of the velar lobes of the foot. Then they press the egg with the velar lobes. The egg collapses due to pressure. Then the eggs are gradually swallowed which do not appear to take more than 2 minutes. In 1946 Thorson reported the process of nurse feeding but this appears different from those of *Nucella* and *Buccinum* types.

References

Annanadale, N. and Kemp, 1916. *Mem. Ind. Mus., Calcutta*, 5: 328–366.

Chari, V.K., 1950. *J. Bombay Nat. Hist. Soc.*, 49: 317.

Chidambaram, K. and Unny, M., 1947. *Proc. Zool. Soc., London*, 117 (2/3): 528–532.

Gravely, 1942. Shells and other animal remains found on the Madras Beach. II. Snails. *Bull. Madras Govt. Mus.*, p. 62–63.

Hornell, J., 1914. The sacred chank of India. A monograph of the Indian Conch (*Turbinella pyrum*). *Mad. Fish. Bull.*, 7: 1–181.

Hornell, J., 1916. Report to the Government of Baroda on the Marine Zoology of Okhamandal in Kattiawar. 2: 1–78.

Hornell, J., 1922. The common mollusks of South India, Madras. *Fish. Bull.*, 1, 14: 97–210.

Panikkar, N.K. and Aiyar, R.G., 1939. *Proc. Ind. Acad. Sci.*, 9B: 343–364.

Panikkar, N.K. and Tampi, P.R.S., 1949. *J. Bombay Nat. Hist. Soc.*, 48(3): 598–599.

Risbec, J., 1928. *Bull. Mus. Hist. Nat.*, Paris.

Risbec, J., 1931. *Ann. Inst. Ocean*, 10(2): 23–33.

Risbec, J., 1932. Note sur les measures de Ricinula et de Fasciolaria. *Bull. Soc. Zool. France*, 57: 358.

Risbec, J.1935. *Ibid*, 60(5): 387–417.

Thorson, G., 1946. Medd. Komm. and Danmarks Fiskeri og. Harunders, Serie: Plankton, Bind 4, Nr. 1.

Chapter 28

Protective Role of Ascorbic Acid on the Lead and Cadmium Induced Alteration in Total RBC of the Freshwater Fish, *Channa orientalis* (Schneider)

☆ *V.R. Borane, B.R. Shinde and R.D. Patil*

ABSTRACT

Freshwater fishes, *Channa orientalis* were exposed to chronic dose of $PbCl_2$ and $CdCl_2$ with and without ascorbic acid. Total R.B.C. counts were recorded. Remarkable decrease in total R.B.C. count was observed in Pb and Cd exposed fishes and the impact was more in lead as compare to Cadmium exposed fishes. Fishes were exposed to heavy metals with L-ascorbic acid showed less present variation on the total R.B.C. count. Pre-exposed Fishes to heavy metals showed fast recovery with ascorbic acid as compared to these cured naturally.

Introduction

Trace amounts of metals occur naturally in water, however wastewater from mining, chemical industries and agriculture etc. Fishes are important as food for people around the world. The first line of defense in fish in the physical barrier comprising skin and mucus that contain immuno reactive molecules that is lysosome complement is immunoglobulin. In fish, the route of heavy metals entry is either through gills or mouth, the blood carries to different organ-system. Hematological studies have a valuable diagnostic tool, in, physiological or metabolic alteration.

Lead is non-essential metal affects human health as it leads to mild hyper chronic and microcytic anemia. The biochemical effect of lead is its interferences with heme synthesis leading to hematological damage (Awad, 1997). The prenatal toxic effects of lead on R.B.C. have recently been studied by scanning E. M. (Dey *et al.*, 1999). Cadmium is release from industrial waste is deposited in bottom sedimentation recorded long term contamination reported by (Leontovicova, 2003).

In animals ascorbic acid content in the tissues increases in stress condition during metal toxicosis. Antioxidant property like Vitamin C can play significant role in the treatment of metal. Ascorbic acid indicate positive role in detoxification behave as efficient chelators (Gurer and Eracel, 2000). Hence to assess the role of ascorbic acid on R.B.C. of an experimental fish, *Channa orientalis* after chronic exposure to lead and cadmium.

Materials and Methods

Medium sized freshwater fishes *channa orientalis* (length 6-8 c.m. and weight 25-35g.) were collected from Shivan River from Nandurbar district and acclimatized in laboratory for 8-10 days. The physico-chemical parameters of the water used by methods APHA and AWWA (1985). The fishes were divided into three groups A, B and C. Group A fishes were maintained as a control. The Group B fishes were exposed to $LC_{50/10}$ dose of Pb^{++} (2.867ppm) as $PbCl_2$ Cd^{++} (1.248ppm) as $CdCl_2$ for 45 days, while group C fishes were exposed to respective chronic concentration of heavy metal with 50mg/l. of ascorbic acid for 45 days. Fishes from B groups were divided into two groups after 45 days exposure to heavy metal into D and E groups. Fishes of D groups were allowed to cure naturally while those of E groups were exposed to ascorbic acid. Total RBC count were recorded from A, B and C group fishes after 15, 30 and 45 days of exposure and from D and E groups after 50'th and 55'th days of recovery.

Blood was obtained by cutting the caudal peduncle dissection method (Roberts 1978; Reichenbach-Klinke, 1982) using heparin as anticoagulant. First few drop were discarded and only the first 2ml. of blood was taken since the entry of lymph into the blood is reported to affect haematocrit value reported by (Schercher 1954). RBC was counted by Neubauer haemocytometer using Hayem's and Tuerk's solution respectively.

Results and Discussion

The data obtained regarding the physico-chemical properties is given in Table 28.1, while the tRBC counts after exposure to lead and cadmium with and without ascorbic acid and during recovery result are given in the Tables 28.1. After chronic exposures to $PbCl_2$ tRBC counts remarkable decreases as compare to $CdCl_2$ were found. In the presence of ascorbic acid (50 mg/l.) tRBC counts decreases less as compared to those of heavy metal intoxicated fishes. The fishes pre-exposed to heavy metals salts showed fast recovery in the RBC counts in the presence of ascorbic acid than those allowed to cure naturally. The treatment with lead and cadmium decrease RBC as compared to control. Whereas, supplementation of ascorbic acid to lead intoxicated exhibits reducing influence of toxicity. Significant decrease in RBC as a result of lead and cadmium intoxicated might be due to hemolytic anemia under the influence of toxic metal.

Jezierska and Witeska (2001) reported that, mucus secreted in large amounts by metal affected gills reduced membrane transport of oxygen in the respiratory epithelium and that disturbed the oxygen diffusion. Drastichová (2004) reported that cadmium was classified among toxic substance for fish. Changes in the R.B.C. profile suggest compensation of oxygen deficit in the

Table 28.1: Physico-chemical Parameters of Eater

Temperature	*25.1±3.2°*
pH	7.6 0±0.3
Free CO_2	3.34±1.3 ml^{-1}
Dissolved O_2	6.3±1.1 ml^{-1}
Total Hardness	204±12.0 mg^{-1}

body due to gill damage. Das and Kumar (2008) reported the treatment of both sublethal and lethal concentration of sevin caused statistically significant and severe reduction in different blood parameters. A decreased HC+, Hb and tRBC in *Aphanius dispar* after Hg exposure has been attributed to the permeability of R.B.C. to mercurial compounds and the balance of the number of binding sites between intracellular extracellular components of blood (Shah and Altindag, 2004). Vitamin C not only confer protection against lead toxicity, but it can also perform therapeutic role against such toxicity in general (Bhattacharjee *et al.*, 2003).

Table 28.2: R.B.C. Counts in *Channa orientalis* after Chronic Exposure to $PbCl_2$ and $CdCl_2$ without and with Ascorbic Acid (millions/mm³)

Group	Treatment	15d	30d	45d	50d	55
A	Control	3.3±0.125	3.2±0.17	3.00±0.294	–	–
B	Pb^{++} (2.867ppm)	2.3±0.057** (–30.30)	2.00±0.21* (–37.5)	1.65±0.092* (–45)	–	–
C	Pb^{++} (2.867ppm)+AA	2.63±0.037** (–20.30)	2.5±0.081* (–21.87)	2.13±0.047* (–29)	–	–
D	Recovery in Normal water	–	–	–	1.92±0.026$^{\Delta}$ [+16.36]	2.10±0.014$^{\Delta\Delta}$ [+27.27]
E	Recovery in AA	–	–	–	2.19±0.143$^{\Delta}$ [+37.72]	2.3±0.163$^{\Delta}$ [+39.39]
B	Cd^{++} (1.248ppm)	2.6±0.139* (–21.21)	2.3±0.129* (–28.12)	1.91±0.081* (–36.33)	–	–
C	Cd^{++} (1.248ppm)+AA	2.71±0.130* (–17.87)	2.61±0.082* (–18.43)	2.4±0.065* (–20)	–	–
D	Recovery in Normal water	–	–	–	2.29±0.018$^{\Delta\Delta}$ [+19.89]	2.50±0.015$^{\Delta\Delta}$ [+30.89]
E	Recovery in AA	–	–	–	2.49±0.129$^{\Delta}$ [+30.36]	2.70±0.02$^{\Delta\Delta\Delta}$ [+41.36]

AA: Ascorbic acid (50 mg/l), ± indicates S.D. of three observations.

Values in ( ) indicates percent change over respective control.

Values in [ ] indicates percent change over 45 days of respective B.

*: Indicates significance with the respective control.

Δ: Indicates significance with 45 days of respective B.

$p < 0.05$ = * and $^{\Delta}$, $p<0.01$ = ** and $^{\Delta\Delta}$, $p<0.001$ = *** and $^{\Delta\Delta\Delta}$, ^{NS} and $^{D\,NS}$ = Not significant.

References

APHA, AWWA and WPCF, 1985. *Standard Methods for the Examination of Water and Wastewater*, 17th edn. APHA Inc., New York.

Das, Tulika and Kumar, Vijay, 2008. Changes in hematological make up of obligate air breathing fish Anabas testudies (Bloch.) exposed to sevin. *Proceeding Zoological Soc., India*, 7(1): 131–133.

Dey, S., Arjun, J. and Das, M., 1999. Erythrocyte membrane dynamics in albino mice offspring born to female with inducted lead toxicity during pregnancy. A scanning electron microscopic study. *Biomed. Lett.*, 59: 55–66.

Drastichová, J., Svobodová, Z., Lusková, V. and Machová, J., 2004. Effect of cadmium on haematological indices of common carp *Cyprinus carpio* L. *Bull. Environ. Contam. Toxicol.*, 72: 725–732.

Gurer, H. and Eracel, N., 2000. Can antioxidant be beneficial in the treatment of lead poisoning? *Free Rad. and Med.*, 29(10): 927–945.

Jezierska, B. and Witeska, M., 2001. Metal toxicity to fish Wydawnictwo, AP sidlece.

Leontovicov'a, D., 2003. Complex monitoring in selected profiles of state networks of water quality follow up in CHMU Periodicum fakulty ekologie a enviromentalistiky Technickej university Vozvolene Vol. 10 Suppl.

Reichenbach-Klinke, 1982. Enfermedades de los peces. Ed Acribia, Zaragoza, España, p. 507.

Roberts, R.J., 1978. *Fish Pathology*. Bailliere Tindall. New York, USA, p. 377.

Schermer, S., 1954. Die Blutmorphologie der laboratorium-stiere Barth, Leipzig. Experta Medica Foundation, FA Davis co. Philadelphia.

Shah, S.L.and Altindag, A., 2004. Haematological parameters of Tech. (*Tinca tinca* L.) after acute and chronic exposure to lethal and sublethal mercury treatments. *Bull. Environ. Contam. Toxicol.*, 73: 911–918.

Chapter 29

Seasonal Variation in Faecal Coliform and Total Coliform Bacterial Density at Pamba River in Comparison with River Achencovil

☆ *Firozia Naseema Jalal and M.G. Sanal Kumar*

ABSTRACT

Four segment of Pamba River were selected to study the physico-chemical and bacterial density of water. The pH, temperature, dissolved oxygen, dissolved carbon dioxide, total dissolved sediments, total coliform and faecal coliform bacterial density were estimated seasonally. Triveni segment of Pamba showed most pollution compared to other three study segments. The bacterial contamination was more in the post monsoon season in all the studied segments. Temperature was low at Triveni in all the studied seasons and pH was neutral. Slight alkaline pH was determined in other segments in all the seasons. Dissolved oxygen and carbon dioxide level of water was almost same in all sites in all the seasons. There is much higher density of faecal coliform and total coliform bacterial density in Pamba river compared with Achencovil river.

Introduction

Pamba River is the third longest river in the state of Kerala and the longest river in the erstwhile princely state of Travancore with a length of 176 km (110 mi). Pamba River has a Source elevation of 1,650 m and average discharge of 109 m^3/s. It covers a basin area of 2,235 km^2 (873).

Pamba originates at Pulachimalai hill in the Peerumedu plateau in the Western Ghats at an altitude of 1650 meters and flows through Ranni, Pathanamthitta, Thiruvalla, Chengannur, Kuttanad and Ambalappuzha Taluks and finally empties into the Vembanad Lake. This river enriches the Pathanamthitta, Alappuzha districts of Kerala state. The Pamba basin is bounded on the east by the Western Ghats. The river shares its northern boundary with the Manimala River basin, while it shares the southern boundary with the Achankovil River basin.

Sabarimala temple dedicated to lord Ayyappa is located near the river Pamba. Sabarimala is one of the major pilgrimage centers of Kerala. The Sabarimala pilgrimage season is from December to February every year. Thousands to one lakhs devotees are coming to this center every day during season. Pilgrims use the water of Pampa for various sanitary purposes.

The water of the River, Pamba is influenced by the wastewater from the pilgrim centre Sabarimala in the upper reaches of the river including the place, Pamba, where the pilgrims arrive, the discharge of wastewater from Municipalities in the middle and lower reaches of the Pamba, forestry and farming especially, the application of fertilizers and pesticides used in plantations and other sources like rubber factories and further industrial and commercial activities. (CESS 2004).

Achankovil River, Achankovilaru in the vernacular, is formed by the confluence of several small streams originating from the hills of Rishimala, Pasukidamettu and Ramakkalteri in eastern Kerala. This small river, less than 130 km long, passes through Mavelikkara, Thiruvalla and Karthikapally Taluks before joining the Pamba River at Veeyapuram in Alappuzha District. Pandalam, the place associated with Lord Ayyappa, is on the banks of this river.

Materials and Methods

Study Sites

Four regions of Pamba River were selected for the study. They were (*a*) Triveni, (*b*) Aranmula, (*c*) Kallissery and (*d*) Veeyapuram. Monthly samples were collected from these study sites. Ten samples were collected from each study sites monthly and the data was changed to seasonal data considering January, February and March as summer; April, May, June as pre monsoon; July, August, September and October as monsoon and November, December as post monsoon.

Water samples were collected in 100ml pre-sterilized glass bottles. Microbial analysis was done by multiple tube fermentation technique (APHA, 2004). Coliform density was calculated as per MPN table.

Secondary data on the faecal coliform and total coliform in various seasons of Achencovil river is obtained from Ecological reports of Achencovil (Sanal Kumar, 2006) for comparison.

Result and Discussion

Bacterial Density

Among the four study regions Veeyapuram showed low total coliform and faecal coliform density in all the studied seasons. Triveni is the site near Sabarimala where anthropogenic disturbance is more. Triveni showed highest total coliform and faecal coliform density in all the studied seasons. This site showed highest number of faecal colifrom during post monsoon season and least during monsoon season.

The study sites Kallissery and Aranmula showed moderate coliform density in all the studied seasons. Among four seasons monsoon season showed lowest coliform density in the study sites and post monsoon season showed highest bacterial density. The density of total coliform and faecal coliform was very low in Achencovil River when compared to Pamba River in all seasons (Table 29.1), (Table 29.2) and (Table 29.3).

Table 29.1: Mean Total Coliform Bacteria (MPN/100ml) at Different Sites in Different Seasons

Sites	*Pre-Monsoon*	*Monsoon*	*Post Monsoon*	*Summer*
Triveni	1186	980	1763	1211
Aranmula	960	600	1110	986
Kallissery	840	480	980	864
Veeyappuram	400	280	460	377

Table 29.2: Mean Faecal Coliform Count (MPN/100ml) at Different Sites in Different Seasons

Sites	*Pre-Monsoon*	*Monsoon*	*Post Monsoon*	*Summer*
Triveni	127	100	136	124
Aranmula	96	98	100	84
Kallissery	86	84	110	60
Veeyappuram	52	41	68	56

Table 29.3: Mean Total Coliform and Faecal Coliform Bacteria (MPN/100ml) Density at Various Sites of Achencovil River in Different Seasons

Site	*Pre-Monsoon*		*Monsoon*		*Post Monsoon*		*Summer*	
	Faecal Coliform	*Total Coliform*	*Faecal Coliform*	*Total Coliform*	*Faecal Coliform*	*Total Coliform*	*Faecal Coliform*	*Total Coliform*
Achencovil (Origin)	4	328	3	410	3	128	4	342
Aruvappulam	6	410	8	600	16	643	14	472
Kollakadavu	12	446	14	348	17	436	32	482
Veeyapuram	16	310	31	380	33	472	32	432

Source: Project report on Ecological studies on river Achencovil, 2006.

Physico-chemical Features of Water

The dissolved oxygen level was very high in all the study sites through out the study seasons. The dissolved carbon dioxide value slightly changed in Kallissery and Veeyapuram during monsoon and summer from other seasons. pH of the water was neutral at Triveni in all the seasons and slightly alkaline at Kallissery, Aranmula and Veeyapuram in all the seasons.

The temperature showed drastic difference at all the sites in all the seasons. The total dissolved solids ranged from 80-700 mg/L in Aranmula, Kallissery and Veeyapuram and between 40-400 mg/L at Triveni during Pre-monsoon, Post monsoon and summer where as it ranged on 100-800 mg/L in monsoon at all the study sites. (Table 29. 4).

Table 29.4: Physico-Chemical Features of Water at Various Study Sites at Different Seasons (Mean Value)

Parameters	*Aranmula*				*Kallissery*				*Veeyapuram*				*Triveni*			
	Pm	*M*	*PoM*	*Su*	*Pm*	*M*	*PoM*	*Su*	*Pm*	*M*	*PoM*	*Su*	*Pm*	*M*	*PoM*	*Su*
Dissolved Cxygen ml/L	6.0	6.1	6.2	6.1	6.1	6.4	6.3	6.2	6.3	6.6	6.4	6.3	6.3	6.8	6.7	6.6
Dissolved Carbon dioxide ml/L	1.11	1.11	1.11	1.11	1.11	1.21	1.10	1.28	1.21	1.38	1.21	1.28	0.09	0.90	0.81	1.20
pH	7.2	7.21	7.00	7.1	7.2	7.21	7.3	7.2	7.12	7.11	7.31	7.2	7.00	7.00	7.00	7.00
Temperature (°C)	34.1	34.6	32.1	35	31.6	33.6	32.6	34.6	31.1	34.1	32.4	34.3	30.6	30.6	31.0	32.2
Total Dissolved Sediments mg/L	200	210	116	216	210	240	186	300	240	280	141	310	80	100	40	60

Pm: Pre-monsoon; M: Monsoon; PoM: Post monsoon; Su: Summer.

From this study it was revealed that coliform density was very high at Triveni during all the studied seasons. This is due to high anthropogenic disturbances associated with Sabarimala pilgrimage. The lowest bacterial density at Veeyapuram is due to less anthropogenic disturbances. Aranmula and Kallissery also showed almost same water quality of Veeyapuram. Among the different studied seasons, monsoon showed lowest coliform density due to dilution.

When compared with river Achencovil it is clear that Pamba River is more polluted. This is due to the reason that only a small portion of Pamba is flowing through forest area and most part is flowing through densily populated villages and towns so that anthropogenic disturbances is more.

The Physico-chemical features of river also showed moderate to adequate levels of parameters like dissolved oxygen, dissolved carbon dioxide, total dissolved sediments etc. The marked temperature difference in four season showed high anthropogenic disturbances expects Veeyapuram.

References

APHA, 2004. *Methods of Water and Wastewater Analysis*. New York, 1822 p.

Badge, H. and Verma, M.C., 1991. Assessment of drinking water quality from a lake. *J. Env. and Poll.*, 5(1): 17–21.

CESS, 2004. Mitigation of Pamba river pollution options strategies and responsibilities.

Mayer, R.K. and Swayer, C.N., 1979. Microbial population. *J. Ecobiol.*, 5(2): 85–88.

Rheinheimer, K., 1971. *Microbial Contamination and Pollution*. SPB Pub., The Hague, 378 pp.

Sanal Kumar, M.G., 2006. Ecological studies of river Achencovil. Project report of local river assessment, pp. 35.

Chapter 30

Checklist of Birds of Ghodpeth Reservoir of Bhadrawati Tahsil in Chandrapur District of Maharashtra State

☆ *N.V. Harney, S.R. Sitre, N.S. Wadhave and P.N. Nasare*

ABSTRACT

Ghodpeth reservoir is a beautiful perennial freshwater reservoir located in Ghodpeth village of Bhadrawati tehsil of Chandrapur district in Maharashtra State on Nagpur–Chandrapur state highway. This is a rain fed reservoir having a water spread area of about 20 acres approximately with a depth of about 15 feet in rainy season. The Ghodpeth reservoir is 6 kms. from Bhadrawati and its catchment area contains a variety of flora and fauna.

This reservoir is biodiversity rich ecosystem and harbours a variety of local as well as migratory birds due to abundant food available throughout the year in the form of crustaceans, insects, worms, mollusks as well as aquatic weeds. The reservoir harbours different kinds of aquatic as well as terrestrial weeds in its catchment area which supports a large number of fauna on which the birds thrive very well.

Studies were made throughout the year for preparing a checklist of birds of this beautiful reservoir. The studies show that this reservoir is having a rich avi-fauna having 44 different kinds of bird species recorded in the catchment area.

Introduction

Ghodpeth reservoir is a beautiful freshwater reservoir located on Nagpur-Chandrapur highway 6 km. away from Bhadrawati tehsil of Chandrapur district. It is a totally rain fed reservoir having abundant water during rainy season. This reservoir is having a total water spread area of about 15 feet in the centre in rainy season, and at the periphery the water depth with 3-5 feet approximately. In summer season water level of the reservoir shrinks considerably and the water depth varies from 3-6 feet only at the centre.

This reservoir harbors a number of aquatic weeds in the submerged as well as floating state on which thrive a large number of organisms. Due to abundant food available throughout the year in this reservoir in the form of aquatic crustaceans, insects, mollusks etc. the reservoir always attracts a large number of birds throughout the year. The banks of this reservoir are infested with weed *Impomea aquatica* which provide suitable resting place for the birds. Apart from this the lake periphery is covered with bushes and trees which provide suitable habitat for many migratory as well as resident birds.

The earlier studies on birds were undertaken by investigators like Majumdar (1984) who studied birds from Bastar district, Newton *et al.* (1986) and Ghosal (1995) listed birds of Kanha tiger reserve, Osmatston (1922) studied birds from Pachmarhi, Yardi *et al.* (2004) reported birds from Salim Ali Lake, Aurangabad, Wadekar and Kasambe (2002) studied birds of Pohara-Malkhed forest reserve, while Kulkarni *et al.* (2005) studied birds in and around Nanded City in Maharashtra.

From the literature survey, it is revealed that this region of Vidarbha is lagging behind in bird studies with respect to various reservoirs. In this context in order to assess how many different kinds of birds are visiting the place throughout the year the present investigation was undertaken to prepare a checklist of avifaunal diversity of Ghodpeth reservoir and its surrounding areas. This investigation will form a base line information resource for future in depth studies on birds of this region of Vidarbha.

Materials and Methods

Ghodpeth reservoir and its surrounding area was visited and surveyed weekly during the study period throughout the year. The birds were observed from safe distance without disturbing them using a field binocular (7x–25x magnification) during morning (6–7 A.M.) and in the evening (6–7 P.M.).

Identification of bird species was done using the field guide given by Salim Ali (2002), Salim Ali and S.D.Ripley (1995) and Grimmett *et al.* (1999). The scientific and local names were ascertained based on the key of Manakadan and Pittie (2001) and status of each species is categorized as Resident (R), Migratory (M) and Resident Migratory (RM). A checklist is finally prepared as per the guidelines given in checklist of birds (Abdulali, 1981, Gaikwad *et al.*, 1997).

Results and Discussion

During the present investigation 39 different bird species including aquatic and non aquatic birds were recorded from the Ghodpeth Reservoir (Table 30.1). It was observed that the maximum bird species were recorded during spring, early monsoon and late winter season, while comparatively less number of bird species were observed during late summer, late rainy season and early winter. The lake harbours a large number of fauna which attracts the migratory as well as non migratory birds which shows that the entire lake basin is highly productive and conducive to all kinds of birds.

In the Present investigation the recorded birds belong to 11 different orders *viz.* Podicipediformes, Anseriformes, Ciconiformes, Falconiformes, Galliformes, Pelecaniformes, Charadriformes, Coumbiformes, Psittaciformes, Passeriformes, and Coraciformes. Table 30.1 gives details about the scientific and common names, occurrence and status.

Table 30.1: Check List of Birds of Ghodpeth Reservoir

Common Name/Order/Family	Scientific Name	Habit
Order–Podicipediformes		
Family Podicipedidae		
Little grebe	*Tachybaptus ruficollius*	R
Order–Ciconiformes		
Family–Ardeidae		
Grey Heron	*Ardea cinerea*	RM
Indian Pond Heron	*Ardeola grayii*	R
Cattle Egret	*Bubulcus ibis*	RM
Large Egret	*Casmerodius albus*	RM
Family–Ciconidae		
Asian Open Bill Stork	*Anastomus Osciatans*	R
Black Necked Stork	*Ciconia episcopus*	M
Family–Threskiornithidae		
Black Ibis	*Pseudibis papillosa*	R
Order–Anseriformes		
Family–Anatidae		
Spot Bill Duck	*Anas poecilorhyncha*	RM
Order–Falconiformes		
Family Accipitridae		
Black Winged Kite	*Elanus caeruleus*	R
Black Kite	*Milvus migrans*	R
Order–Galliformes		
Family–Phasinidae		
Grey Francolin	*Francolinus pondicerianus*	R
Order–Gruiformes		
Family–Gruidae		
White Breasted Water Hen	*Amaurormi Phoenicurus*	R
Purple Moorhen	*Porphyrio porphyrio*	R
Common Coot	*Fulica atra*	RM
Order–Pelecaniformes		
Family–Phalcrocoracidae		
Little Cormorant	*Phalacrocorax niger*	RM
Order–Charadriformes		
Family–Recurvirostridae		
Black Winged Stilt	*Himantopus himantopus*	*R*
Family–Charadridae		
Red Wattled Lapwing	*Vanellus indicus*	*R*
Family–Scolopacidae		
Common Sandpiper	*Actitis hyoleucos*	M

Contd...

Table 30.1–Contd...

Common Name/Order/Family	Scientific Name	Habit
Order–Columbiformes		
Family–Columbidae		
Little Brown Dove	*Streptopelia senegalensis*	R
Order–Psittaciformes		
Family–Psittacidae		
Rose Ringed Parakeet	*Psittacula krameri*	R
Family–Cuculidae		
Asian Koel	*Eudynamis scolopacea*	R
Greater Coucal	*Centropus sinensis*	R
Order–Coracifores		
Family–Alcedinidae		
Small Blue Kingfisher	*Alcedo athis*	RM
White Breasted Kingfisher	*Halcyon smyrnesis*	R
Family–Meropidae		
Small Green Bee Eater	*Merops orientalis*	R
Family–Coraciidae		
Indian Roller	*Coracias benghalensis*	RM
Family–Upupidae		
Common Hoopoe	*Eupopa epops*	RM
Order–Passeriformes		
Family–Lanidae		
Rufousbacked shrike	*Lanius schach*	R
Family–Dicrudidae		
Black drongo	*Dicrurus macrocercusq*	R
Family–Sturnidae		
Common myna	*Acridotheres tristis*	R
Brahminy starling	*Sturnus pagodarum*	R
Family–Pycnonotidae		
Red Vented bulbul	*Pycnonotus cafer*	R
Family–Muscicapidae		
Jungle babbler	*Turdoides striatus*	R
Indian Robin	*Saxicolodies fulicata*	R
Family–Necatarinidae		
Purple Sunbird	*Nectarinia asiatica*	R
Family–Passeridae		
House sparrow	*Passer Domesticus*	R
Pheasant Tailed Jackana	*Hydrophasianus chirurgus*	R
Order–Passeriformes		
Family–Hirudinidae		
Common Swallow	*Hirundo rustica*	RM

R: Resident; RM: Resident Migratory; M: Migratory.

The little grebe, Common sand piper, Little Cormorant and Black Winged Stilt were observed in October. Common Sandpiper, Rufosubacked shrike, Brahminy starling and Purple moorhen in November, Asian Open bill stork, Black winged kite in February, Shikra was recorded in March month. The Brahminy starling was recorded in Winter. Black necked stork a threatened species was also recorded during winter months in the reservoir basin. The overall checklist prepared show that 39 different kinds of birds have visited the reservoir water for feeding and breeding activities during the year, as abundant food available in the reservoir water.

References

Ali, S. and Ripley, S.D., 1983. *A Pictorial Guide to the Birds of the Indian Sub-continent*. Bombay Natural History Society (BNHS), Mumbai, pp. 1–354.

Ali, S., 2002. *The Book of Indian Birds*. Bombay Natural History Society, Mumbai, pp. 1–354.

Ghosal, D.K., 1995. Avi fauna of Conservation Areas, No. 7, Fauna of Kanha Tiger Reserve, Zoological Survey of India (ZSI), pp. 63–91.

Grimmette, K., Inskipp, C. and Inskipp, T., 1999. *Birds of Indian Sub Continent*. Oxford University Press, New Delhi, pp. 1–384.

Kulkarni, A.N., Kanwate, V.S. and Deshpande, V.D., 2005. Birds in and around Nanded City, Maharashtra. *A Zoos Print Journal*, 20(11): 2076–2078.

Majumdar, N., 1984. On a collection of birds from Bastar District, M.P. *Record Zoological Survey of India, Occasional Paper*, 59: 54.

Manakadan, R. and Pittie, A., 2001. Standardized common and Scientific names of Birds of the Indian Subcontinent. *Bueros*, 6(1): 1–37.

Newton, P.N., Brudin, S. and Guy, J., 1986. The birds of Kanha Tiger Reserve, Madhya Pradesh, India. *J. Bombay Natural History Society*, 83(3): 477–498.

Osmaston, B.B., 1922a. Birds of Panchmarhi. *J. Bombay Nat. Hist. Society*, 83(3): 477–498.

Wadatkar, J.S. and Kasambe, R., 2002. Checklistof birds from Pohara–Malkhed Reserve Forest, Dist. Amravati, Maharashtra. *Zoos Print Journal*, 17(66): 807–811.

Chapter 31

Impact of Sugar Mill Wastewater on Groundwater Quality

☆ *R.D. Joshi and S.S. Patil*

ABSTRACT

The sugar industry wastewater is responsible for severe pollution problems like air, water and soil pollution there by affecting adversely the drinking water quality of various dug-wells bore wells in general. Therefore, an attempt has been made in this paper to study the effect of sugar mill wastewater on the groundwater quality of Ambajogai sugar factory area. In the studies, it was observed that since there were no proper disposal methodologies employed for treatment of distillery waste, the groundwater reservoirs such as dug wells were contaminated through groundwater percolation of distillery wastewaters. The well waters were contaminated with distillery wastewaters thereby affecting its physiochemical characters adversely, resulting in non potability of these water reservoirs.

Introduction

Maharashtra is one of the leading state with more than 200 sugar factories owned by private and cooperative sectors. In the recent 15 years *i.e.* 1990 to 2005 many sugar factories established distilleries as subsidiary industries to minimize the pollution problem as well for extra financial gain through alcohol production and selling. But this in turn leads to severe environmental pollution problems for soil and water.

This in spite of the fact that these agrobase industries are the backbone of rural economy of our state, subsequently an urgent need has arisen to review and take a serious note of the other associated problems as like soil, air and water pollution. The disposal of wastewater arising from these industries with or without proper treatment has been a great cause of constant deterioration of groundwater quality, thus affecting water supply for agriculture and drinking purpose. The impact of groundwater pollution due to percolation of groundwater thus, is prominent. Further, the use of spentwash for irrigation purpose has helped to spread the pollution in soil. The color and odor of groundwater observed in some parts of the study area has aesthetically made groundwater to unfit for drinking purpose. The concentration of metals like Fe, Mn, in few samples is also above the maximum permissible limit for drinking water. The groundwater has also found to be unsuitable for the irrigation purpose, as the EC values of majority of samples falls in very high salinity zone (Pondhe, 2005). Thus, the problem of wastewater pollution from sugar mill wastewater in general and distilleries in particular is becoming more and more severe. There are nearly 200 distilleries providing 150 million lits of alcohol and releasing agent 20,000 million lit of spentwash every (Saraf, 1988). It is estimated that the groundwater generated in each distillery is 12-15 lit per lit of alcohol provided *i.e.* a distillery producing 40,000 lit of alcohol/day discharges 5-6 lakh lit of spentwash (Tare, 1980).

Since no proper and judicial implementation is there on the methods of sugar mill waste disposal, our state is posing/facing a serious problem of soil and water pollution resulting in affecting the productivity of soil adversely with groundwater pollution.

Taking all this into account the present investigation was undertaken to assess the effect of sugar mill effluent on groundwater quality of Ambajogai area, in the Beed dist of Maharashtra state.

Well inventory surveys was carried out by collecting well water samples in affected (within 5 km) and unaffected (5 km away) with their physico-chemical analysis with seasonal variation was carried our for one year *i.e.* Oct. 2006 to Sep. 2007. Ten well water samples, and sugar mill wastewater, along with spentwash were collected and analysed for physicochemical characters as color, pH, COD, BOD, Na, K, P, TS, sulfate, sulfide, TDS, DO, nitrates, acidity, electrical conductivity, hardness, alkalinity, chlorides etc.

Material and Methods

Groundwater samples from dug wells were obtained after pumping the well for ½ hr. The samples were then properly labeled, with the record of depth of well depth of water table and operating condition of the wells. The sugar mill wastes and distillery wastewaters were collected in suitable containes as they come out from the factories. The samples were labeled properly and were analysed with important physico-chemical parameters (Trivedy/Goel, APHA-AWWA 1975). The collected water, and effluent samples were brought to the laboratory. The analysis were carried out with the help of suitable laboratory analytical or instrumental methods available.

The collected groundwater samples were brought to laboratory to carry out the chemical analysis of the constituents. These samples were filtered through Whatman No.42 filter paper. Chemical analysis on filtered water samples was carried out to estimate/determine pH, Electrical conductivity (EC), Sodium (Na^{2+}), Potassium (K^{+}), Calcium (Ca^{2+}), Magnesium (Mg^{2+}) Chloride (Cl^{-}), Total alkalinity (as $CaCO_3$), Total hardness (as $CaCO_3$), Sulfate (SO_4^{2-}), Phosphate (PO_4^{3-}), Nitrate (NO_3), Dissolved Oxygen (DO), Chemical Oxygen Demand (COD), Biochemical Oxygen Demand (BOD) etc. The methods used for this analysis were standardized according to the procedure given in standard method or examination of water and wastewater by APHA-AWWA and WPCF (1975; Trivedy and Goel, 1986; Gaines, 1993; Kodarkar *et al.*, 1998). The results are reported in parts per million (ppm).

Results and Discussion

Characteristics of Sugar Mill Wastes in Ambajogai

Distillery wastewater, spentwash is a very hot liquid as it is released from the distillery. It has a dark intense brown colour, acidic pH 3.8. With high level of COD, BOD values and rich in other organic and inorganic constituents as shown in the table.

In Ambajogai sugar factory and distillery area both the wastes are released without any physicochemical treatment.

Mill house wastes from Ambajogai sugar factory is 840 lit/day/tone of cane. The boiling house waste, filter cake washings, condenser water waste from Ambajogai sugar factory were 330, 380 and 2020 lit/day/tone of cane sugar.

Tables 31.1 and 31.2 shows that the sugar mill waste and distillery waste characteristics from Ambajogai sugar factory sugar mill waste has slightly alkaline pH (8.3), with TDS (3920), TS (110), Chlorides (552), Sulfates (620) ppm. The distillery waste from the sugar factory shows acidic pH, (3.8), High BOD, COD values as 72000 ppm and 151720 ppm respectively. Similarly, the spentwash also contains higher load of TS, TDS, as 12.332 gm/lit and 3.72 gm/lit, and sulfate sulfide contents as 4900 ppm and 32 ppm respectively. Moreover, the intense brown color due to melanoidin pigment with off obnoxious flavor makes the soil and water environment polluted.

Table 31.1: Volumes of Different Effluents from Various Areas of Ambajogai Sugar Factory

Sl.No.	*Source of Effluent*	*Average Quantity lit/day/tone Cane*
1.	Mill House	840
2.	Boiling House	330
3.	Filter cloth washings	380
4.	Condenser water	2020

Table 31.2: Characterization of Distillery Wastewater, Spentwash from Ambajogai Sugar Factory

	Distillery Waste (Ambajogai)
Colour	Dark brown
pH	3.8
COD	151720 ppm
BOD	72000 ppm
Na	0.28 per cent
TKN	0.21 per cent
P	0.29 per cent
TS	12.332 gm/lit
TVS	3.72 gm/lit
Sulfate	4900 ppm
Sulfide	32 ppm

Higher BOD, COD values (Table 31.2) are observed in distillery wastes of Ambajogai sugar factory. Similarly TS, TDS, sulfates and sulfide contents are also significantly more in the distillery waste of Ambajogai sugar factory. These values indicate that the pollution load caused by the total wastes (sugar mill waste + spentwash) from Ambajogai sugar factory is severe. No proper pretreatment methodologies, were employed for the sugar mill waste in general before its disposal in the study areas.

In Ambajogai sugar factory, the average pH for one year ranged from 6.9 to 7.43 in the sugar industry. During season the pH ranges from 6.9 to maximum 7.55.

The average EC for one year is 590–2826.33 in the sugar factory season months (Oct 06–May 07) it ranges from 530 to 2934.5. Well No. 1 2, shows high EC since these well were unlined.

The average range of hardness throughout the study year was 65.33 to 418.33. However during the season of the sugar factory *i.e.* Oct 06 to May 07 it ranged from 62-417.5. Well No. 1 and 2 showed relatively high values of Hardness since these wells were unlined.

The average DO for one year is 2.6 to 7.3 mg/lit but during the season of the sugar factory it was between 2.9 to 7.9.

The BOD value for one year is Oct 06 to Sept 07 studies indicated the range from 5.3 to 8.5667 mg/lit while in the season of the sugar factory (Oct 06 to May 07) it elevates to maximum 9.25 mg/lit. The

Table 31.3: Average Physico-Chemical Data for the Groundwater from the Study Area of Ambajogai Sugar Factory for One Year (October 06 to September 07)

Sl. No.	Depth of Well (in mtrs)	Depth of Water Table (in mtrs)	pH	EC	TDS	TH	Major Constituents							Minor Constituents		Indicator Parameters		
							Cationic				Anionic							
							Na	K	Ca	Mg	Cl	TA	SO_4	Po_4^{3-}	No_3^-	DO	BOD	COD
1.	25	14	7.033	2826.33	1650	418.33	556.66	7.1	170	19.333	1226.66	400	199.333	0.95333	2.81	7.3	8.0667	26.267
2.	30	11.66	7.1	2545.33	1656.66	385	508.33	7.033	151.666	22.666	979.33	439.333	197.333	0.80333	2.16	5.4	8.5667	24.467
3.	32	14	7.56	723.33	566.66	76.66	117.33	5.066	37.333	10.666	227	153.666	47.333	0.42	0.72333	2.8333	5.3	14.933
4.	18	10.66	7.33	516.66	396	80.33	88.33	4.8	53.666	10.666	244.33	136	52	0.31667	0.42	3.0	4.5333	14.033
5.	22	13.66	6.96	590	437.33	65.33	82.66	3.1333	58	13.333	227.66	129.333	62	0.42	0.49666	2.6	3.6333	12.133
6.	32	15.33	7.43	595	421.66	69.66	89.33	5.1	37.666	12	227.66	122	53.666	0.53333	0.4	5.2333	5.4	13.733
7.	27	13	7.16	730	499.3	62.33	70	5.466	31.666	14.333	239	129.333	60	0.57333	0.46333	4.7	4.9	14.1
8.	35	13.66	7.3	796.66	610.66	95	71.33	5.866	37.333	11.666	180.10	94.333	42.333	0.27666	0.24333	3.9666	5.5	14.633
9.	32	13.33	6.9	683.33	551.66	94	70.66	4.66	37.333	15	195	87.333	43	0.19	0.26033	4.4	5.6333	14.8
10.	25	10.33	7.26	710	476.66	81	74.66	3.9	43.33	17.33	208.33	112.333	55.666	0.24666	0.48333	4.6666	183	15

All the values are in mg/lit except pH and EC.

Table 31.4: Average Physico-Chemical Data for the Groundwater from the Study Area of Ambajogai Sugar Factory During the Season (October 06 to May 07)

Sl. No.	Depth of Well (in mtrs)	Depth of Water Table (in mtrs)	pH	EC	TDS	TH	Major Constituents							Minor Constituents		Indicator Parameters		
							Cationic				Anionic							
							Na	K	Ca	Mg	Cl	TA	SO_4	Po_4^{3-}	No_3^-	DO	BOD	COD
1.	25	11.5	7.15	2934.5	1635	417.5	575	6.2	165	13.5	1330	405	199	0.95	2.665	7.9	8.35	26.15
2.	30	8.5	7.4	2813	1695	397.5	572.5	7.3	145	15	879	474	201	0.795	1.79	6.05	9.25	24.2
3.	32	11.5	7.85	825	615	80	121	6	38.5	12.5	128	167.5	47.5	0.47	0.775	3.2	5.35	14.4
4.	18	8	7.25	530	399	86.5	87.5	5.5	61.5	10.5	137.5	138	54	0.325	0.33	3.5	4.35	13.35
5.	22	15	7.2	530	396	62	68	3.6	72	11	82	110	65	0.42	0.41	2.9	3	10.5
6.	32	14.5	7.55	552.5	392.5	67	75	5.4	30.5	7	78.5	97	49.5	0.56	0.25	6.8	5.5	12.5
7.	27	11	6.95	825	544	71	72.5	6.8	28.5	11.5	98.5	109	56	0.6	0.32	5.9	4.65	12.75
8.	35	12	7.05	922.5	707.5	112.5	69.5	7.2	38.5	9	35.16	72.5	37.5	0.185	0.02	4.8	5.55	13.5
9.	32	10.5	6.95	785	635	108.5	66	5.15	33.5	13	56.5	60	36.5	0.045	0.0455	5.45	5.75	13.75
10.	25	10	6.95	760	485	87.5	68.5	3.75	38.5	14.5	67.5	93.5	54.5	0.12	0.375	5.95	271.9	14.35

All the values are in mg/lit except pH and EC.

wells which are unlined and near to the factory are showed high BOD while these wells which are away and lined show minimum or low BOD values.

The average COD values for one year were 14.1 to maximum of 26.267. The seasonal months of sugar factory *i.e.* Oct 06 to May 07 showed the COD values 10.5 to 26.15 mg/lit. The high COD values in well No. 1 and 2 are due to unlining of the wells resulting in more percolation of waste organic substances from sugar mill waste into wells. The average alkalinity observed throughout for one year was 87.333 to 439.333 maximum, while in the sugar factory seasonal month *i.e.* Oct 06 to May 07 it ranged as 60 to 474 mg/lit.

The average values for one year *i.e.* Oct 06 to Sept 07 ranged from 180.10 to 1226.66, while those values in the seasonal months of the sugar factory are 56.5 to 1330.00 mg/lit. The average sulphate contents for one year studies ranged from 43 mg/lit to 199.333 mg/lit, while in the seasonal month *i.e.* Oct 06 to May 07 these values were 36.5 to maximum 201.00 mg/lit.

The average of phosphate contents through out for one year *i.e.* Oct 06 to Sept. 07 were 0.19 to maximum 0.953 mg/lit whereas during the seasonal months of the sugar factory these values were 0.6 to 0.795 maximum. The nitrate contents for one year *i.e.* Oct 06 to Sept. 07. 0.4 to 2.81. The studies on cationic constituent of groundwater samples of 10 dug wells around Ambajogai sugar factory area are given below.

Na–70–556.66 mg/lit for one year, K–3.9 to 7.033 mg/lit for one year, Ca–31.666–170.00 mg/lit for one year, Mg–12.00–22.666 mg/lit for one year. The values for the seasonal months ranges from 68-575 mg/lit (Na), 3.6–7.3 (K), 28.5–165.0 (Ca) and 7.0 to 14.5 mg/lit (Mg).

All the above values for the physico-chemical analysis of sugar mill waste and spentwash indicated that both these wastes are definitely containing biotoxic constituents including high BOD and COD values. The dark brown color, acidic pH is definitely inhibitory for microbiota as well leading to groundwater pollution. The survey for Ten dug well samples for one year also indicated that the groundwater is definitely getting contaminated with the sugar mill and distillery wastes. The analysis showed high concentration of TDS, cationic, anioic substances. The pH, EC, TH is also higher. The indicator parameter values such as COD, BOD, and DO are also quite higher and toxic to aquatic life. The groundwater reservoirs within the 5km area around the sugar factory is getting polluted with the sugar industry wastes. Thus, these water reservoirs are becoming useless not only for agricultural use but also for the drinking purpose. Since, the primary sources of drinking and agricultural water are dug wells contaminations in the wells by sugar industry wastes is causing a severe threat. Almost all of our observations resembles with the studies performed by Alam (2008) Ansari (2001) and Pondhe (2005). The impact of sugar mill effluent quality on the groundwater studies reveals the similarities with the studies performed by Rafeeque (2002) and Khan (2002). Thus, it can be concluded that the sugar mill wastewaters are definitely inhibitory for the surrounding environment in general, leading to groundwater pollution in particular in the Ambajogai sugar factory area. This may be due to improper pretreatment methods employed for the effluent treatment.

References

Alam, M., Kumar, M., Thakur, S.K. and Jha, C.K., 2008. Effluent of distillery effluent on soil, crop and groundwater. *Indian Sugar*, p. 47–50.

Ansari, J.A., Jalukar, R.S. and Malik, G.M., 2001. Performance evaluation of effluent treatment plant of a sugar industry. *Oriental J. of Chem.*, 17(1): 115–117.

Barnhan, A.K., Sharma, R.N. and Borch, G.C., 1993. Impact of sugar mill and distillery wastewater on water quality of river Gelabi Assam. *Indian J. of Env. Health*, 35(4): 288–293.

Deshpande Jayshree and Lomte, V.S., 2007. Impact of distillery effluent on water and soil quality. In: *Sustainable Environmental Management*. Daya Publishing House, New Delhi, p. 52–63.

Kumar, A., Singh, Y., Joshi, B.D. and Rai, J.P.N., 2003. Effect of distillery spentwash on some characteristic of soil and water. *Indian J. Ecol.*, 30. 7–12.

Pandey, A.K., Sidiqui, S.Z. and Ramrao, K.V., 1993. Physico-chemical and biological characteristics of Hussan sagar an industrially polluted lake, Hyderabad. *Rec. Acad. Env. Biol.*, 2(2): 101–107.

Pondhe, G.M., 2005. Impact of agrobase industries on the groundwater quality of Sonai area dist Ahamednagar. *Ph.D. Thesis* in Env. Sci.

Rafeeque, M.A. and Khan, A.M., 2002. Impact of sugar mill effluent on the quality of river Godavari near Kandakurthi village, Nizamabad dist A.P. *J. Aquat. Biol.*, 17(2): 33–35.

Saraf, R.V., 1988. Spentwash: A new source for production of chemicals. In: *Proceedings of National Seminar on Sugar Factory and Allied Industries Wastes: A New Focus*, p. 63–71.

Tare, Vinod, 1988. Treatment and disposal of distillery waste in pollution management industries. R.K. Trivedy, p. 1–13.

Trivedy, R.K. and Goel, P.K., 1986. *Chemical and Biological Methods for Water Pollution Studies*. Environ. Publ., Karad, India, 251 pp.

Chapter 32

Physico-chemical Properties of Groundwater of Jintur Taluka in Parbhani District of Maharashtra

☆ *V.B. Pawar and Kshama Khobragade*

Introduction

Unplanned urbanization, rapid industrialization and indiscriminate use of artificial chemical in agriculture, causing heavy and varied pollution in aquatic environment leading to deterioration of water quality depletion of aquatic biota. The knowing to understand chemical content of water it is difficult to biological phenomenon fully, because chemical nature of water, reveals much about the metabolism of the ecosystem and explain the general hydro biological interrelationship.

The water which moves down word percolates in soil becomes groundwater further reacts with the soil and rock materials. These reactions primarily consist of solution of solid phase in accordance with the solution of chemistry of particular minerals. These minerals range from almost insoluble to very soluble. The solubility also affected by temperature and pressure.

Various workers in our country have carried out extensive studies on fluoride Das *et al.* (2000) reported fluoride hazardous of groundwater in Orissa, Khedkar and Dixit (2003), evaluated suitability of Ambanala water for Irrigation. Recently Kumar and Gopal (2000) published a review article on fluorosis and its preventive strategies. Chand (1999) reported fluoride study on human health and Patra. Diwedi and *et al.,* Studied industrial fluorosis on cattle in Udaipur, Rajasthan. Jagdap *et al.* (2002); Hussain *et al.* (2003) and Abbasi *et al.* (2002), have studied water quality in different rivers. Sriniwas *et al.* (2000) and Jha *et al.* (2000) and Warma (2003) studied water quality in Hyderabad and Bihar respectively Patnaik area. A survey of literature reveals that there is no systematic study on evaluation of Physico-chemical properties in groundwater of Jintur Taluka of Parbhani District. Hence present work has been undertaken for the study.

Materials and Methods

The study area lies in the Parbhani District of Maharashtra State. Jintur is located at 19.62° N 76.7° E. It has an average elevation of 455 meters (1492 feet). Water samples are collected from 15 numbers of selected Bore wells as well as of open well in sterilized bottles (Kudesia, 1985) of one liter capacity during October 2007 to February 2008. Water samples were collected in morning at 8.00 A.M. to 11 A.M. Sampling has been carried out without adding any preservative in rinsed bottles directly for avoiding any contamination and brought to the laboratory. Only high pure chemicals and double distilled water used for preparing solutions for analysis. Bacteriological examination was done using standard procedure suggested by Trivedy and Goel (1984) and manual of APHA (1989).

Various physical parameters like pH, EC, DO and TDS which are important to evaluate the suitability of groundwater for portability were determined on the site with the help of digital portable water analyzer kit (CENTURY-CK-710). The chemical analysis was carried out for Calcium (Ca^{2+}), Magnesium (Mg^{2+}), Chloride (Cl^-), Carbonate (CO_3^{2-}) and Bicarbonate (HCO^-_3) by volumetric litigation methods, while fluoride (F-) by spectrophotometer method. Sodium (Na^+) and Potassium (K^+) are by flamephotometery (Kudesia, 1985).

Result and Discussions

The characteristics of groundwater collected from different sites are presented in Table 32.1. the pH values of water samples are varying from 6.2 to 8.28 and these values within the limits except samples S8 (8.28), prescribed by ISI and ICMR but all of these samples within the limits as per WHO (1993) Table 32.3 gives the Physico-chemical properties of groundwater of Jintur Taluka of Parbhani District.

Table 32.1: Sampling Stations of groundwater of Jintur taluka of Parbhani District.

Sl.No.	Station Code	Name of Stations
1.	S1	Adarsh Colony
2.	S2	Anand Nagar
3.	S3	Nivriti Mohalla
4.	S4	Lecture colony
5.	S5	Bori village
6.	S6	Bamani
7.	S7	Bhogaon
8.	S8	Itoli
9.	S9	Jamb
10.	S10	Yeldari
11.	S11	Limbala
12.	S12	Pungala
13.	S13	Bori Tanda
14.	S14	Wazar
15.	S15	Warud

Conductivity values varied from 300 to 2986μ siemens/cm for groundwater and these values are very much higher than the prescribed standards limits (1400μ siemens/cm) recommended by WHO except S5, S6 and S10 (Figure 32.1) higher EC and TDS values reflect greater salinity of water and it can not be suitable for drinking and irrigation under ordinary conditions but may be used occasionally under special circumstances. By using the only EC values (Table 32.2) Wilcox (1995) has classified the groundwater for irrigational waters. By considering this classification more than 50 per cent of samples of study area were found to be doubtful to unsuitable and the classification is as follows.

Total Dissolved Solids (TDS) values varied from 120 to 1525 mg/liter Figure 32.2 except S2, S9, S13, S14 and S15 remaining all sampling station TDS values are higher than the standard values recommended by ISI, ICMR and WHO. The samples which have high value of TDS are unsuitable for Drinking and Irrigation purpose.

Table 32.2: Classification of Samples According to their Class of Samples

EC (μ siemens/cm)	Class	No. of Samples
<250	Excellent	–
250-750	Good	5
750-2000	Permissible	7
2000-3000	Doubtful	1
>3000	Unsuitable	2

Table 32.3: Physico-chemical Properties of Groundwater of Jintur Area of Parbhani District

Sampling Station No.	Location	pH	Turbidity	EC	TDS	DO	TA	Ca^{2+}	Mg_2^+	Cl_2^-	FI	E. coli
1	Adarsh Colony	7.86	5.0	1887*	960	5.1	226	195	65	225	0.20	–
2	Anand Nagar	7.91	4.2	535	370	4.8	186	210*	70	220	0.29	–
3	Nivratti Mohallah	6.64	3.1	969	610	4.6	140	280*	95	124	0.16	1
4	Lecture colony	7.62	3.6	1578*	946	5.8	205	485*	164*	226	0.05	–
5	Bori	7.06	3.2	2748*	1525	4.9	210	385*	185*	180	0.12	2*
6	Bamni	7.5	2.8	2986	874	5.9	246	615*	134	139	0.16	–
7	Bhogaon	6.2	5.3	1395	530	4.6	198	198	65	136	0.28	–
8	Itoli	8.28	6.2	934	510	5.1	230	172	69	121	0.53	–
9	Jamb	6.92	5.6	958	120	4.8	138	126	69	85	0.43	2*
10	Yeldari	7.04	5.3	2542*	1210	4.5	102	340*	185*	160	0.16	1
11	Limbala	7.58	5.0	1162	520	5.1	120	176	93	114	0.29	14*
12	Pungla	7.19	8.0	680	210	4.8	44	56	35	24	0.16	120*
13	Bori Tanda	7.3	6.2	480	160	4.6	78	52	26	28	2.43	81*
14	Wazar	6.4	6.1	300	120	5.1	90	68	42	15	0.15	45*
15	Warud	6.87	6.2	690	150	4.9	172	143	113	40	0.13	19*

*: Exceeding the permissible limit

All Parameters are Expressed in Mg/lit. Except pH,Turbidity (NTU), Electrical Conductivity (μ mhos/cm) and *E. coli* (Number/100 ml.)

Hardness of water relates to its reactions with soap and to scale and incrustations accumulating in conduct where the water is heated or transported. Since soap if precipitated by Ca and Mg ions. Water classified according to its hardness as:

Soft	0-60
Moderately Hard	61-120
Hard	121-180
Very Hard	>181

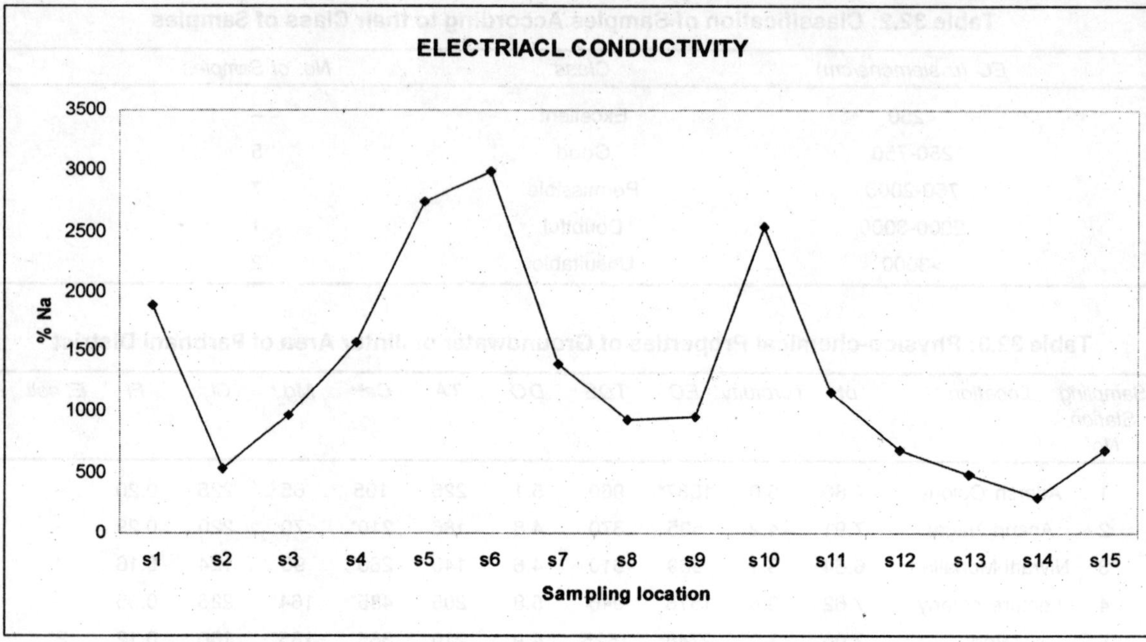

Figure 32.1: Variation of EC in Groundwater of Jintur Taluka of Parbhani District

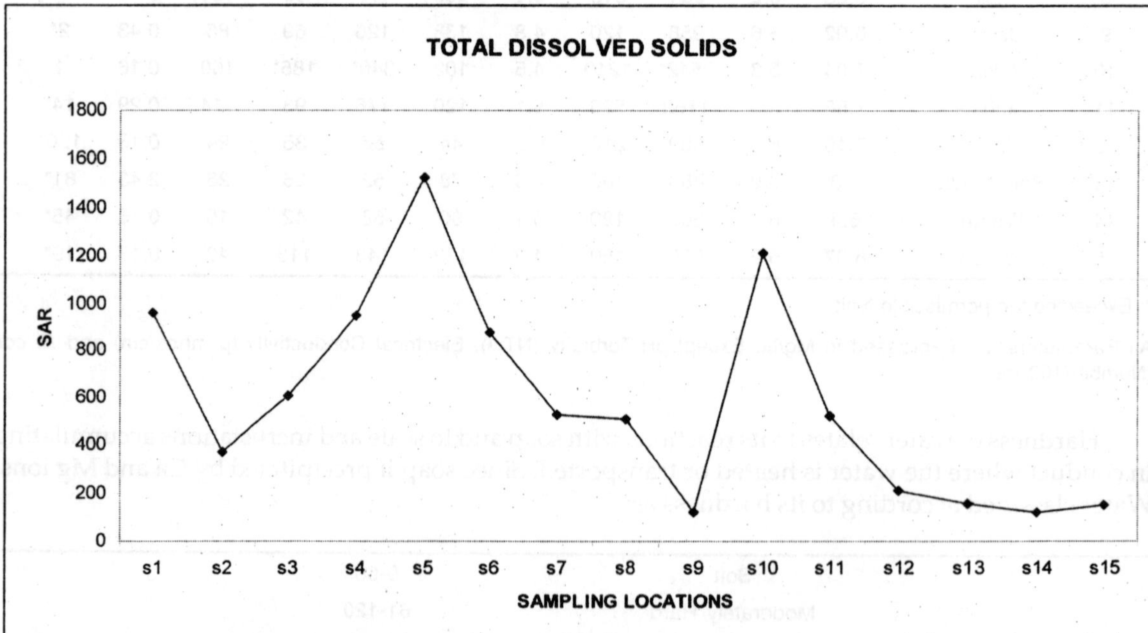

Figure 32.2: Variation of TDS in Groundwater of Jintur Taluka of Parbhani District

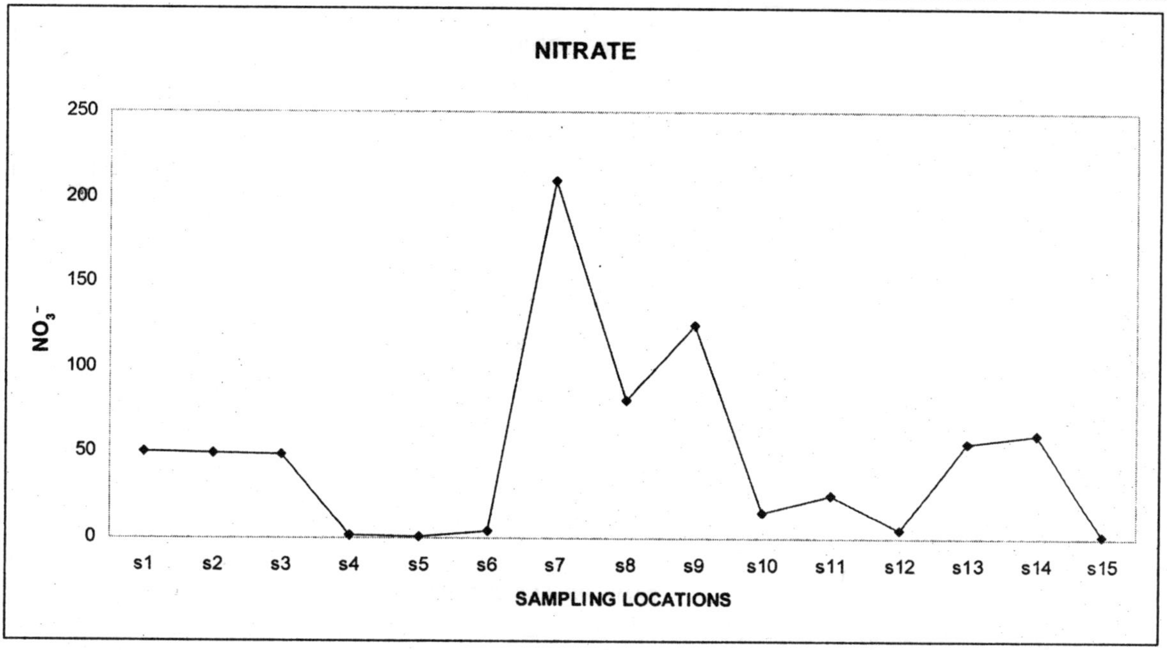

Figure 32.3: Variation of NO^{3-} in Groundwater of Jintur Taluka of Parbhani District

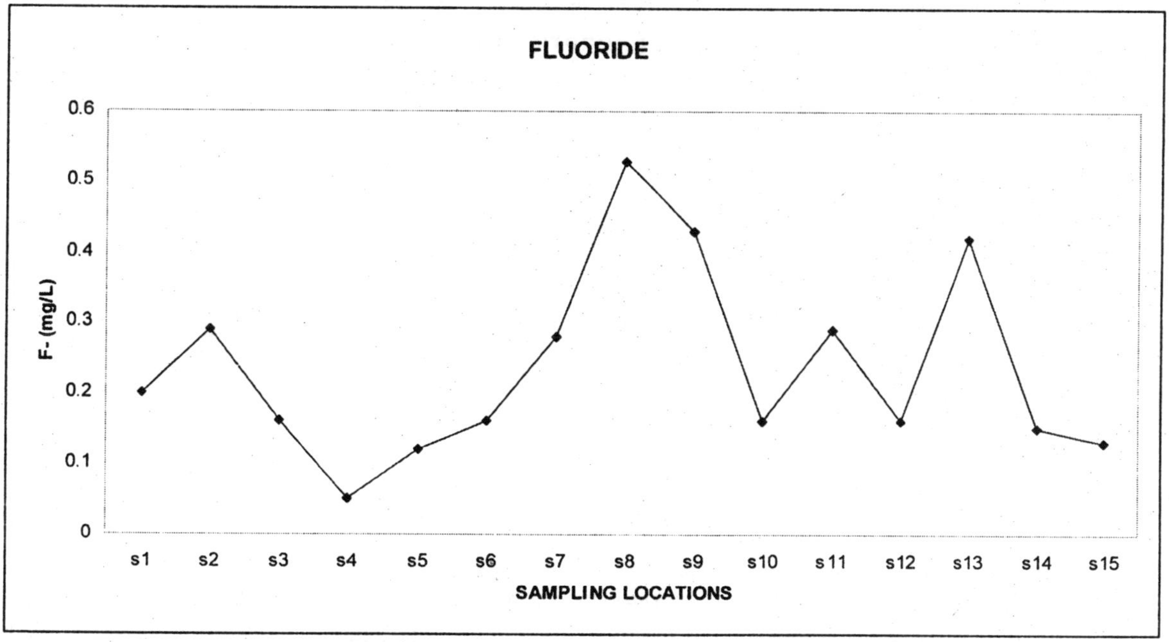

Figure 32.4: Variation of F^- in Groundwater of Jintur Taluka of Parbhani District

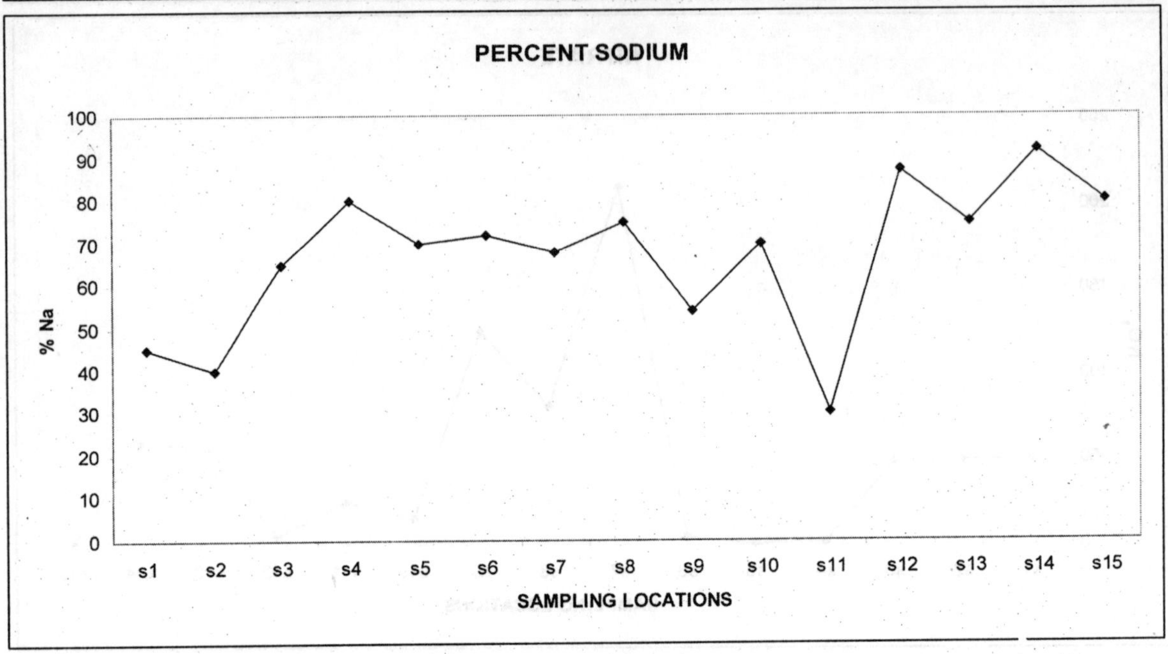

Figure 32.5: Variation of Per cent Na in Groundwater of Jintur Taluka of Parbhani District

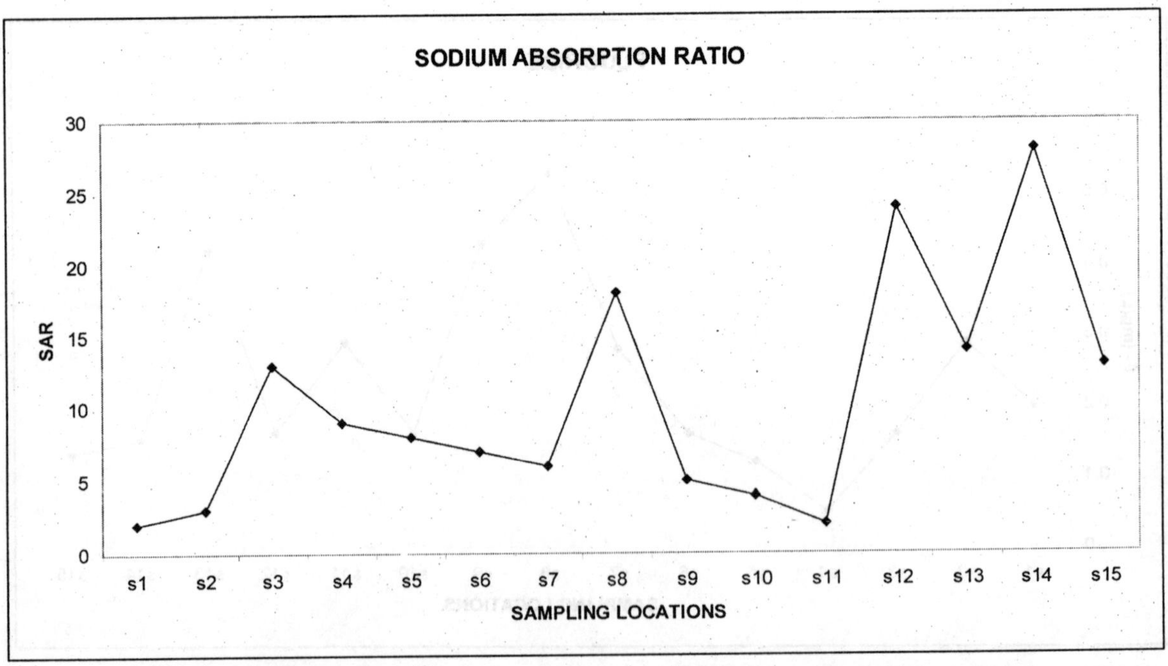

Figure 32.6: Variation of SAR in Groundwater of Jintur Taluka of Parbhani District

Water for domestic use should not contain more than 80 mg/liter total hardness, Hardness of groundwater can be increased if contaminated by acid leachate from garbage waste disposal areas. The disadvantage of hard water is that it precipitate soap thus increasing soap requirements low incidence of heart diseases apparently also occur in areas with very soft water which led Neri *et al.* (1975).

Potassium values ranging from 0.89 to 24 mg/liter Potassium is common in igneous rock. These minerals however are very insoluble so that potassium levels in groundwater normally are much lower than sodium concentration.

Sodium Absorption Ratio is defined as following formula

$$SAR = Na^{+}/\sqrt{[(Ca^{2+}+Mg^{2+})/2]}$$

Where the individual ions have been expressed as meq/L. there is a significant correlation between SAR values of irrigation water and the extent to which sodium is absorbed by the soil. The SAR values from 1.35 to 28.42 (Figure 32.6) in which samples S3, S8, S12, S13, S14 and S15 have high value of SAR recommended by USSL, so these samples are not suitable for irrigation purpose.

Bicarbonate concentration of water has been suggested as an additional tool for classification of irrigation of water by Richard (APHA, 1989). Bicarbonate hazard is evaluated by calculating residual sodium carbonate (RSC) and based on these values groundwater of the investigated area is suffer from bicarbonate hazard and the classification is as follows.

RSC Values	*Class*	*No. of Samples*
<1.25	Good	11
1.25-2.5	Medium	01
>2.5	Bad	03

Calcium hardness values are varied from 52 to 615 mg/L and these values are within permissible limit as prescribed by ICMR and WHO. Sources of Calcium are igneous rock minerals like silicate pyroxenes, amphiboles feldspar and silicate minerals produced in metamorphism. Since the solubility of these minerals is low in calcium as well as in TDS.

Magnesium hardness values are varied from 26.0 to 185 mg/L and are within permissible limit as prescribed by ICMR and WHO standards except S4, S5 and S10 magnesium in around water from igneous rock primarily derives from ferromagnesian minerals like Olivine, pyroxenes, amphiboles.

Chlorides values varied from 15 to 226 mg/L and these values are within permissible limits as per prescribed by ICMR, WHO standards except S1 and S4. In reasonable concentration these are not harmful to human beings impart salty taste beyond concentration 250 mg/L which is objectionable to many peoples.

Sulphate values are varied from zero to 210 mg/L and are within permissible limit as prescribed by ISI, ICMR and WHO drinking water should be not exceed 250 mg/L of Sulphate because water will have a better taste.

Carbonates and Bicarbonates values from 04 to 113.0 mg/L and 77 to 397 mg/L respectively all of the studied samples Bicarbonate values are lesser than the prescribed by US standards and carbonates has no standard values for drinking purpose suggested ISI (1991), ICMR(1975) and WHO. Sources of Carbonate and Bicarbonate include CO_2 from atmosphere CO_2 Sulphate produced by the biota of the soil or by the activity of Sulphate producers and the various carbonate rocks and minerals.

Nitrate values varied from 05 to 215 mg/L all the studies samples nitrate value are lesser than the prescribed by ICMR, WHO standards except S7, S9 and S14 (Figure 32.3). Nitrate occurs naturally in certain water supplies and may also find access through directly or indirectly discharges of wastes and sewage.

Fluoride values of fifteen samples varies from 0.05 to 2.43 mg/l except samples S13 which are varied from 2.43 mg/L, so the samples are higher than the prescribed limit recommended by ICMR, ISI and WHO. These groundwater very hazardous for human consumption.

The DO was recorded in the range of 4.5 to 5.9 to mg/L the highest value of DO was encountered in the samples S6. There were not any variations of DO values found and they all are in Permissible limit. All of the samples have lower BOD values except S10. The higher BOD indicates pollution from domestic sources.

Conclusion

The general conclusion can be drawn as the quality of almost all groundwater samples collected from the study area indicate that the concentration of pH, Ca^{2+}, Mg^{2+}, SO^{2-}, DO, BOD and F values are within permissible of ISI, ICMR and WHO but Cl^-, TDS, NO^-_3 and EC values are that poor water quality in most of the studied groundwater samples. A study area classifies water under moderate category and is not best for house hold and irrigation.

References

Abbasi, S.A., Khan, F.I., Senthilveal, K. and Shobudeen, A., 2002. Modelling of Buckinggham canal water quality. *Indian J. Environ. Health*, 44(4): 290–297.

APHA, 1989. *Standard Methods for Examination of Water and Wastewater*, 17th Edn., (Ed.) Lenore S. Clescrei. APHA, AWWA, WPCE, Washington DC.

Chand, D., 1999. Fluoride and human health: Causes for concern. *Indian Journal of Environmental Protection*, 19(2): 81–89.

Das, S., Mehta, B.C., Samanta, S.K., Das, P.K. and Srivastava, S.K., 2000. *Indian Journal of Environmental Health*, 1(1): 40–46.

Hussain, M.F. and Ahmad, I., 2003. Variability in physico-chemical parameters of Pachin River (Hanagar). *Indian J. Emv. Health*, 44(4): 329–336.

Jayshri, Jagdap, Bhushan, Kachawe, Leena, Deshpande and Prakash, Kelkar, 2002. Water quality assessment of the purna river for irrigation purpose in Buldhana district, Maharashtra. *Indian J. Environ. Health*, 44(3): 247–257.

Jha, A.N. and Verma, P.K., 2000. Physico-chemical property of drinking water in town area of Godda District under Santal Pargana, Bihar (India). *Pollution Research*, 19(2): 245–247.

Kumar, S. and Gopal, K., 2000. *Indian Journal of Environment Protection*, 20(6): 430.

Patnaik, K.N., Satyanarayan, S.V. and Swoyam, Poor Rout, 2002. Water pollution from major industries in Pradeep area: A case study. *Indian Journal Environment Health*, 44(3): 203–211.

Sriniwas, C.H., Tisko, R.S., Venkateshwar, C., Satnarayan Rao, M.S. and Reddy Ravindra, R., 2000. Studies on groundwater quality of Hyderabad. *Poll. Res.*, 19(2): 285–289.

Sharma, Surendera Kumar and Chandel, C. P. Singh, 2006. Physico-chemical properties of groundwater of Dudo block of Jaipur District. *Ecology, Environment and Conservation*, 12(1): 141–147.

Chpater 33

Marine Algae Hemagglutinins from the Coast of Goa

☆ *Kumar Sudhir, Tiwary Mukesh, Kumari Switi and Barros Urmila*

ABSTRACT

Aqueous extract from nine Indian marine algae were examined for hemagglutination activity using native human blood cells, the crude extracts from five species revealed hemagglutinating activity against tested human blood cells, whereas the results obtained did not indicate specificity towards any particular human blood cells. Further aqueous extracts from four species were subjected to saturation F/0-90 ammonium sulphate precipitation. The precipitated products were assessed for titer activity and any blood group specificity. To date, this is the first report of hemagglutinins extracted from marine algae, collected from the shores of Goa, India.

Introduction

Lectins constitute a group of proteins or glycoproteins of non-immune origin, which bind reversibly to carbohydrates and thus combine with glycocomponents on the cell surface. They are widely distributed in nature and are reported to be present randomly in plants, algae, fungi, microorganisms, viruses and animals (vertebrates and invertebrates). Due to their chemical properties, lectins play a pivotal role in cell–cell recognition and thus consequently find application in the fields of immunology, membrane structure studies, cell biology and cancer research to name a few. However, there is limited data/information about algal lectins in comparison with that from higher plants and vertebrates.

The first report on the presence of lectins in algae was made by Boyd *et al.* (1966) in 24 algal species from the coast of Puerto Rico. Although there are several reports on the occurrence of agglutinins in marine algae (Britain Blunden *et al.*, 1978, Japan Hori *et al.*, 1988, Spain Fabregas *et al.*, 1992, South East United States Bird *et al.*, 1993, Brazil Freitas *et al.*, 1997 and Vietnam Dinh *et al.*, 2009).

The aims of the present study deals with crude as well as ammonium sulphate saturated extract from screened marine algae having high hemagglutinating activity against native human blood cells, which indicates the presence of lectin.

Materials and Methods

Sample Collection

Marine macroalgae abundantly found along the coast of Goa, India, was collected during the months September to January, with reference to the lowest tides. The algae were cleaned of epiphytes, rinsed with distilled water and stored at –20°C until use.

Extraction of Crude

The method employed for crude extraction (modification of Sampaio *et al.*, 2002) of hemagglutinins using Phosphate buffered saline at pH 7.4. The frozen algae were grounded in mortar and pestle and homogenized with (1: 3 w/v) of Phosphate Buffered Saline, for 18 hrs while stirring at 4°C. The homogenate was filtered through muslin cloth, centrifuged at 12,000 rpm at 4°C for 30 minutes and the supernatant obtained was stored at –20°C until further use.

Ammonium Sulphate Precipitation of Crude Extracts

The precipitation of crude extract experiment was carried out at 4°C. Solid Ammonium Sulphate was added slowly to crude extract. The precipitate formed at 0-90 per cent, the precipitate was separated by centrifugation at 12000 rpm at 4°C for 30 minutes. The pellet was dissolved in minimal volume of extracted buffers, and dialyzed against Phosphate Buffered Saline with gradual mixing at 4°C. Every six hrs, the PBS was changed for fresh buffer and dialysis continued for another 24 hrs.

Preparation of RBC Suspension

Approximately 1 ml of human erythrocyte was centrifuged at 3000 rpm for 5 min at room temperature. The supernatant of serum was discarded and erythrocyte pellets was washed with cold PBS. It was then centrifuged and finally 2 per cent of erythrocytes suspension was prepared for hemagglutination assay.

Assay for Hemagglutination

Hemagglutination assays for different algal crude extract were conducted on human erythrocyte suspensions in 96-well microtitre plate. 75 µl of different algal extract was mixed with 75 µl of 2 per cent erythrocyte suspension. The control used as positive control was erythrocytes with antisera of respective blood groups. Negative control was erythrocytes with PBS. The micro titer plate was gently shaken and was incubated in a moist chamber for one hour at room temperature. Hemagglutination was recorded by visual observation and by comparing the agglutination with positive and negative controls.

Results

Assay for Hemagglutination

Out of nine macro algae screened five showed hemagglutination activity when tested with human blood cells. The results are tabulated in Table 33.1. *Grateloupia lithophila, Scinaia hateai, Hypnea musciformis, Acanthophora specifera* showed strong hemagglutination activity from crude extract, where as *Gelidium serrulatum* showed moderate activity against tested erythrocytes A^+, O^+. Remaining crude extracts were failed to agglutinate tested human blood cells.

Table 33.1: Hemagglutination Activity from Aqueous Extract

Macroalgae	A^+	B	O^+
G. lithophila	+	+	+
S. hateai	+	+	+
H. musciformis	+	+	+
A. specifera	+	+	+
G. corticata	–	–	–
P. veitnensis	–	–	–
G. serrulatum	+/–	+/–	+/–
H. valentiae	–	–	–
A. fragilissima	–	–	–

+: Good agglutination; +/–: Partial activity; –: No Activity.

Hemagglutination Titer

The hemagglutination titers are reported as the minimum amount of protein extract tested that produced agglutination. The ammonium sulphate saturated extract from *Grateloupia lithophila, Scinaia hateai, Hypnea musciformis* and *Acanthophora specifera* were the most potent agglutinating agents against human blood cells A^+, B^+ and O^+. The partially purified extract from *Grateloupia lithophila* and *Hypnea musciformis* was the most potent agglutinating against tested erythrocytes O^+, rather than A^+, B^+. The results are tabulated in Table 33.2.

Table 33.2: Hemagglutination Titer of Ammonium Sulphate Saturated Extracts Comparing Titer Versus Protein Concentration (µg)

Macroalgae	*Blood Cells*	*Titer Value*	*Protein (µg)*
S. hateai	Blood group A	2^1	18.7
	Blood group B	2^1	18.7
	Blood group O	2^2	8.35
G. lithophila	Blood group A	2^4	10.3
	Blood group B	2^4	10.3
	Blood group O	2^8	5.15
H. musciformis	Blood group A	2^8	8.43
	Blood group B	2^8	8.43
	Blood group O	2^{16}	4.21
A. specifera	Blood group A	2^2	33
	Blood group B	2^2	33
	Blood group O	2^4	16.5

Discussion

Marine macroalgae are widely distributed throughout the world and abundantly present along the coast of Goa, India. This is the first report regarding hemagglutinins from screened macroalgae and report of the hemagglutinins found along the shore of Goa, India. Lectin activity had previously been examined with positive results with at least one of different types of animal and human erythrocytes. Most of the authors have reported that enzyme treated animal erythrocytes have been more suitable for lectins detection from marine algae than human erythrocytes (Hori *et al.*, 1988; Fabregas *et al.*, 1985; Chiles and Bird, 1989; Freitas *et al.*, 1997). From our screening studies, native human erythrocytes were tested to detect the hemagglutination activity from crude extract. Nine different macroalgae were extracted and assessed for hemagglutination activity. Majority of the crude extract exhibited agglutination activity except four, but results obtained did not show any exclusive specificity towards particular blood cells. Further four extract were selected for saturation using ammonium sulphate precipitation F/90. The saturated extracts were assessed for titer activity against tested erythrocytes. Ammonium sulphate saturated extracts showed higher titer 1: 8 to 1: 16 by lower range concentration of proteins. But none of the partially purified products showed specificity towards particular blood cells. Our titer is low, as enzyme-treated or animal erythrocytes always show higher titer (Ainouz *et al.*, 1992; Chiles and Bird 1989). Ammonium sulphate saturated extracts from all four species appeared to be highly specific for human O^+ erythrocytes rather than A^+ and B^+ There is a general trend in which human type A^+ blood is agglutinated more weakly by these algal extracts than types O^+ or B^+ erythrocytes. Maximum studies on marine algal lectins have appeared to be highly specific for human B^+ erythrocytes (Hori *et al.*, 1981; Fabregas *et al.*, 1985; Chiles and Bird, 1989).

These data indicates that collected algae contain strong hemagglutinin with a wide range of differences, suggesting that they could be useful tools in carbohydrate research and lectin-based diagnosis. Due to these interesting results, further investigations of these algal hemagglutinins are under progress.

Acknowledgements

Special thanks to Dr. Urmila Barros and Dr.R.V.Bhagde for their valuable support.

References

Ainouz, I.L., Sampaio, A.H., Benevides, N.M.B., Freitas, A.L.P., Costa, F.H.F., Carvalho, M.C. and Pinheiro, Joventino F., 1992. Agglutination of enzyme treated erythrocytes by Brazilian marine algae. *Bot. Mar.*, 35: 475–479.

Bird, K.T., Chiles, T.C., Longley, R.E., Kendrick, A.F. and Kinkema, M.D., 1993. Agglutinins from marine macroalgae of the southeastern United States. *J. Appl. Phycol.*, 5: 213–218.

Blunden, G., Rogers, D.J. and Farnham, W.F., 1978. Hemagglutinins in British marine algae and their possible taxonomic value. In: *Modern Approaches to the Taxonomy of Red and Brown Algae*, (Eds.) D.E. Irvine and J.H. Price. Academic press, London, pp. 21–45.

Boyd, W.C., Almodovar, L.R. and Boyd, J.G., 1966. Agglutinins in marine algae for human erythrocytes. *Transfusion* (Philadelphia), 6: 82–83.

Chiles, T.C. and Bird, K.T., 1989. A comparative study of the animal erythrocyte agglutinins from marine algae. *Comp. Biochem. Physiol.*, 94B: 107–111.

Dinh, H.L., Hori, K. and Quang, N.H., 2009. Screening and preliminary characterization of hemagglutinins in Vietnamese marine algae. *J. Appl. Phycol.*, 21: 89–97.

Fabregas, J., Llovo, J. and Munoz, A., 1985. Hemagglutinins in red seaweeds. *Bot. Mar.*, 28: 517–520.

Fabregas, J., Lopez, A., Llovo, J. and Munoz, A., 1992. A comparative study of seafish erythrocytes and agglutinins from seaweeds. *Comp. Biochem. Physiol.*, 103A2: 307–313.

Freitas, A.L.P., Teixeira, D.I.A., Costa, F.H.F., Farias, W.R.L., Lobato, A.S.C., Sampaio, A.H. and Benevides, N.M.B., 1997. A new survey of Brazilian marine algae for agglutinins. *J. Appl. Phycol.*, 9: 495–501.

Hori, K., Miyazawa, K. and Ito, K., 1981. Hemagglutinins in marine algae. *Bull. Jpn. Soc. Sci. Fish.*, 47: 793–798.

Hori, K., Oiwa, C., Miyazawa, K. and Ito, K., 1988. Evidence for wide distribution of agglutinins in marine algae. *Bot. Mar.*, 31: 133–138.

Sampaio, A.H., Rogers, D.J., Barwell, C.J., Saker-Sampaio, S., Nascimento, K.S., Nagano, C.S. and Farias, R.L., 2002. New affinity procedure for the isolation and further characterization of blood group B specific lectin from the red marine alga *Ptilota plumosa*. *J. Appl. Phycol.*, 14: 489–495.

Chapter 34

Two New Distributional Records of Bivalve Species of Family Spondylidae from Mandapam Area, South-East Coast of India

☆ *C. Stella, S. Vijayalakshmi and A. Murugan*

ABSTRACT

The new occurrence of Two species of bivalves from Spondylidae family was recorded from Mandapam area based on a few shells collected from the fish landing centers. The two species are under the family of Spondylidae, *Spondlylus longitudinalis and Spondylus obliqus*. The present paper described the taxonomic status and the description of the two species of bivalves of Spondylidae family collected from Mandapam area.

Introduction

Spondylus is a genus of bivalve molluscs, the only genus in the family Spondylidae. As well as being the systematic name, Spondylidae is the most often used common name for these animals, though they are also known as thorny Oysters or Spiny Oysters. There are many species of Spondylus and they vary considerably in appearance and range. They are grouped in the same super family as the Scallops, but like the true Oysters (Family: Ostreidae) they cement themselves to rocks, rather than attaching themselves by a byssus. Their key characteristic is that the two parts of their shell are hinged together with a ball and socket type of hinge, rather than a toothed hinge as is more common in other bivalves (Berrin *et al.*, 1997).

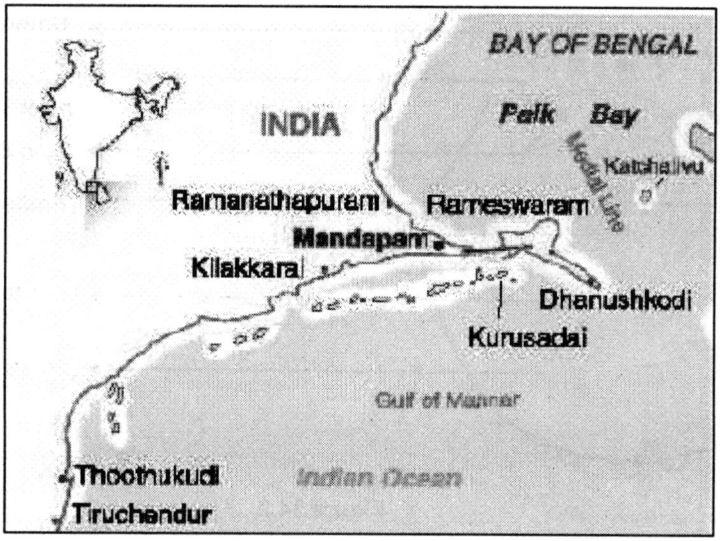

Figure 34.1: Study Area

During the regular survey at Mandapam (Figure 34.1), were collected four or five shells of each species for identification. Identification is mainly based on the external morphology of the shell. The two species are under the family of Spondylidae.

Systematic Position

Phylum	:	Mollusca
Class	:	Bivalvia
Order	:	Ostreoida
Suborder	:	Pectinina
Superfamily	:	Pectinoidea
Family	:	Spondylidae (Gray, 1826)
Genus	:	Spondylus (Linnaeus, 1758)

Family: Spondylidae (Gray, 1826)

Genus: Spondylus (Linnaeus, 1758)

Shell stout, irregularly rounded and higher than long. Outer sculpture mainly radial, often scaly to spinose, umbones separated from hinge line by a triangular area. Hinge line straight with small triangular expansion at each end. Ligament internal in a deep medium pit, single adductor muscle scar, no pallial sinus.

1. *Spondylus obliqus* (Chenu, 1845)

Size: 5.6cm

Locality: Coral area, Intertidal area

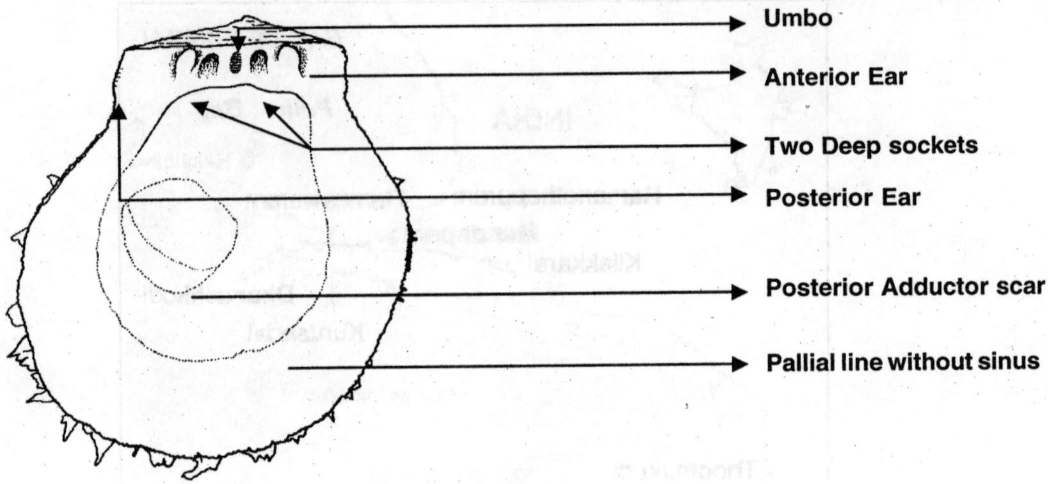

Figure 34.2

Common name: Thorny Oysters

Status: Uncommon

Distribution: Gulf of Mannar: Philippines (35 mm)

Description

Shell upto 5.6cm in length, thick and heavy. Outer surface of the shell rough with small spine. Shell narrow and elongated. Surface consist of numerous concentric rings, thick, elongated narrow radial ribs are present. Reddish brown in colour. Inner surface of the shell smooth and white,edge purple, umbo center, two deep socket present, small ligament, blunt and small teeth present. Posterior adductor scar present and pallial line without sinus (Figure 34.3).

2. *Spondlylus longitudinalis* (Lamarck 1819)

Size: 5 cm

Locality: Coral area, Intertidal area

Common name: Thorny Oysters

Status: Uncommon

Distribution: Gulf of Mannar: Elsewhere: Kyung Nam, Korea (60 mm)

Description

Shell upto 5 cm in length, thick and heavy often blunt ended. Outer surface of the shell rough, minute radial ribs with concentric rings present, numerous small minor cords are present on the radial ribs. Umbo center. Inside the shell white, small ligament, both ears are reducing in nature, 2 deep sockets present, two teeth are present facing upwards, posterior adductor scar present. Pallial line without sinus. Inside the shell edges orange in colour (Figure 34.4).

Figure 34.3: ***Spondylus obligus***

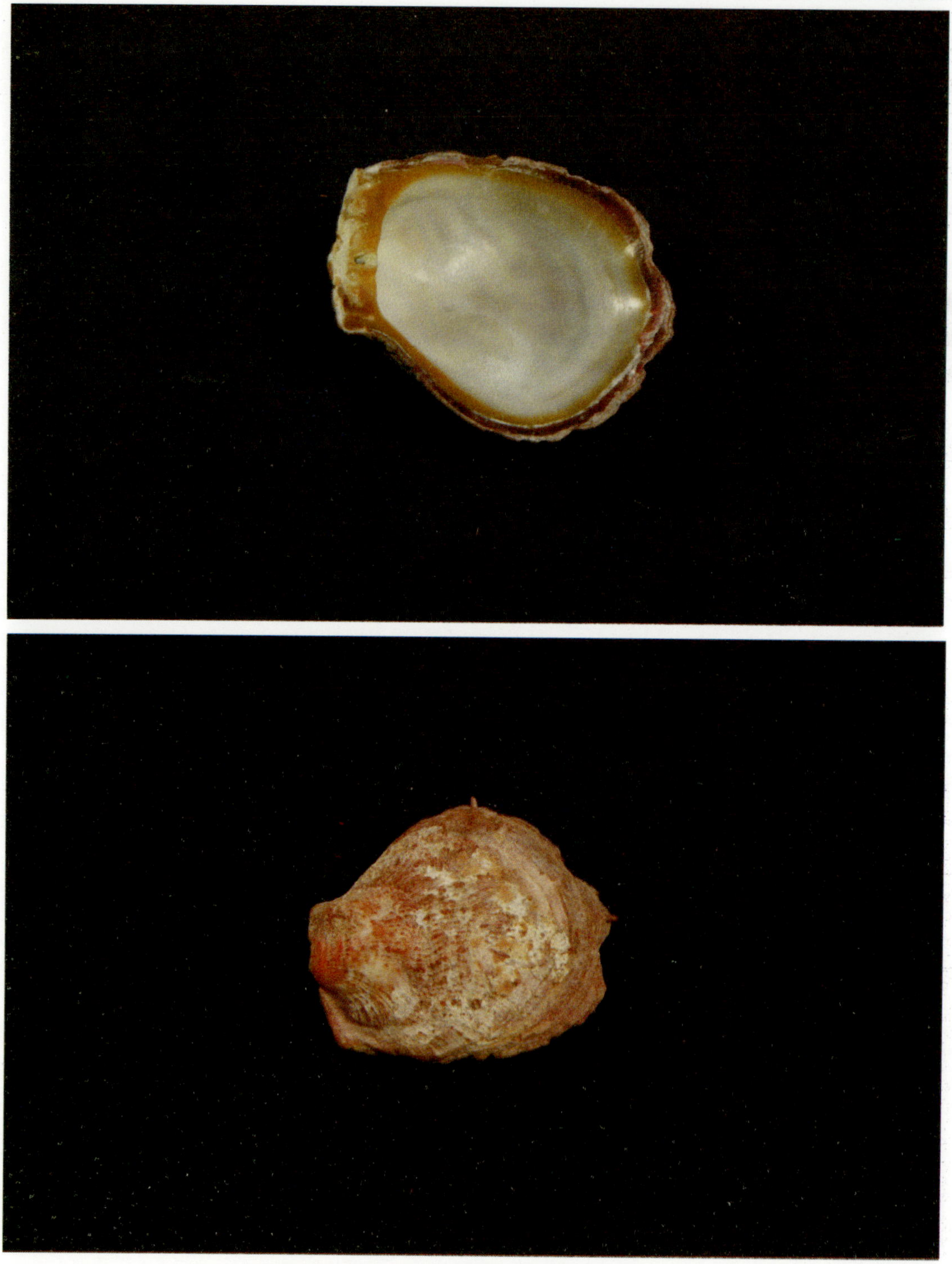

Figure 34.4: ***Spondylus longitudinalis***

Remarks

One specimen ot each species has been deposited in the Zoological survey of India–Chennai (Reg. No. 172, 171). The shells were collected from the fish landing centers of Mandapam. In fish landing centers the shells were collected from the discarding of trawl catches from Gulf of Mannar. Previously these species has been recorded in Kyung Nam, Korea (60) and Philippines (35 mm).

References

Berrin, Katherine and Museum, Larco, 1997. *The Spirit of Ancient Peru: Treasures from the Museo Arqueologico Rafael Larco Herrera.* Thames and Hudson, New York.

http//en.wikipedia.org/wiki/spondylidae.

Chapter 35

Protein Content Variation in Some Body Components of *Barytelphusa guerini* After Exposure to Zinc Sulphate

☆ *R.P. Mali and Shaikh Afsar*

ABSTRACT

The static bioassay test for acute toxicity of $ZnSO_4$ were conducted for 0, 24, 48, 72, 96 and 120 hrs. Various tissue like chelate leg muscle and gill muscle of *Barytelphusa guerini* were analysed for total protein variation. After acute exposure the experimental crabs were subjected to biochemical analysis along with control set. The control animals for leg muscles showed slight decrease up to 48 hrs and sharp decline up to 72 hrs, then it significantly increases up to 96 hrs. The gill muscle activities are similar to that of leg muscle except the initial increase at 24 hrs. The experimental set of both showed slight increase up to 48 hrs and significantly decrease up to 72 hrs, again with sharp increase at 96 hrs and finally fall up the level. By considering the effects of these inorganic pollutants the present study is aimed to determine the sublethal concentration of Zinc sulphate to *B. guerini.*

Introduction

Increasing environmental pollution due to wide use of heavy metals from agricultural, industrial and domestic sectors results release of some toxic metals causes serious impacts to the aquatic animals. Environmental pollutants bring about the damage to different organs or disturb the physiological and biochemical processes. The physiological functions of animals get disturbed on exposure to pollution stress. A particular pollutant incites a specific type of physiological change. The aquatic organisms are susceptible to pollution effect of heavy metals, pesticides as well as industrial efficient. But normally an organism tries to adapt itself to these changes by changing their metabolic activities, but at higher concentration these pollutants can cause damage to the physiological system by affecting the organism either at organ cellular level or even at molecular level. Many people have worked on the effects of different pesticides and heavy metals on the physiology and biochemical aspects of different animals (Mali, 2010; Jagtap, 2009; Sailaja, 2008; Rajaiah, 2007).

Proteins are important organic substances required by an organism in tissues building and repair. They are mostly high molecular weight compounds. Amino acids helps in tonic balance of cellular media and formation of many enzymes. Under extreme stress conditions, proteins have been known to act as the energy supplier in metabolic pathways and biochemical reactions.

Materials and Methods

The crabs *Barytelphusa guerini* used for experimentation, were collection from paddy fields of Nanded district. They were maintained in glass aquarium jars and acclimatized for three days. Only healthy female crabs ranging between 30 to 50 grams were selected for present work to avoid the effect of sex and size (Ambore, 1976). The animals were subjected to sublethal concentration of $ZnSO_4$ and protein estimation was studied after exposure at 0, 24, 48, 72, 96 and 120 hrs.

To study the effect of heavy metal on biochemical changes, the crab was exposed to sublethal concentration 0.02 per cent of Zinc sulphate colorimetric estimation of total proteins was done by using Lowry's method (1951) at 660 nm. The Bovine serum albumin was used as a standard.

Table 1: Mean values of six samples of control and Treated animals in µg/ml

Exposure Periods in hrs	*Control*	*Treated*	*Exposure Periods in hrs*	*Control*	*Treated*
	Leg Muscle Proteins µg/ml	*Leg Muscle Proteins µg/ml*		*Gill Muscles Proteins µg/ml*	*Gill Muscles Proteins µg/ml*
0 Hrs	271	–	0 Hrs	231	–
24 Hrs	180	205.66	24 Hrs	280	264.00
48 Hrs	185	232.33	48 Hrs	225	329.00
72 Hrs	115	120.00	72 Hrs	155	228.33
96 Hrs	205	311.60	96 Hrs	315	325.00
120 Hrs	235	213.30	120 Hrs	325	206.6

Results and Discussion

The results of control and treated animals for leg muscle and gill muscles showed as indicated in above (Table 35.1). The results of control animals for leg muscle show initial decrease up to 48 hrs and sharp decline up to 72 hrs, then it significantly increase up to 96 hrs. Fish, crab and shrimps are the main aquatic organisms in the food chain which may often accumulate large amounts of certain metals (Agoes Soegianto, 2008). The gill muscle activities are similar to that of leg muscle except the initial increase at 24 hrs. Bioaccumulated heavy metals like, mercury, lead, zinc, copper bind with biologically active constituents of the body such as lipids, amino acids, enzymes, and proteins (Passow *et al.*, 1961). The experimental set of both showed slight increase up to 48 hrs and significant decrease up to 72 hrs, again with sharp increase at 96 hrs and finally fall up the level. This up and down activities of treated animal indicates uneasy physiological activities of those animals. Both the endogenous and exogenous factors operate simultaneously to influence the body composition of fish (Haard 1992). Proteins are important biomolecules involved in a wide spectrum of cellular function.

They interplay between enzymatic and non-enzymatic proteins to govern the metabolic harmony (Lehinger 1984).

As a result sometimes the content may surpass even the normal level.Tissue specific and time dependent loss in the protein content was recorded in the tissues of a freshwater field crab *Barytelphusa guerini* after sodium fluoride intoxication (Reddy and Venugopal, 1990). From these observations it can be concluded that the inorganic pollutant like Zinc sulfate can impose a stress on crab, which leads to decrease in protein content initially while afterward the animal tries to regain the normal level to with stand against the stress.

References

Agoes, Soegianto, 2008. Bioaccumulation of heavy metals in some commercial animals caught from selected coastal waters of East Java, Indonesia. *Research Journal of Agriculture and Biological Sciences*, 4(6): 881–885.

Haard, N.F., 1992. Control of chemical composition of food quality attributes of cultured fish. *Food Res. Int.*, 25: 289–307.

Jagtap, A.R., Afsar, S.K., Kothole, S.D. and Mali, R.P., 2009. Effect of pollutant from car washing centre on oxygen consumption in freshwater fish, *Channa punctatus*. *J. Aqu. Biol.*, 24(2): 189–192.

Lowry, O.H., Roserbrough, N.J., Farr, A.L. and Randall, R.J., 1951. Protein measurement with folin phenol. reagent. *J. Biol. Chem.*, 193: 265–275.

Lehinger, A.L., 1984. In: *Biochemistry*, 3rd edn. Kalyani Publisher, Ludhiana, New Delhi.

Mali, R.P., Afsar, S.K. and Jagtap, A.R., 2010. Impact of cadmium induced alterations in the glycolytic potential of freshwater female crab *Barytelphusa guerini*. *Geobios*, 37: 100–102.

Passow, H. and Clarkson, T.W., 1961. The genal pharmacology of heavy metals. *Pharm. Rev.*, 13: 185–224.

Reddy, S.L.N. and Venugopal, N.B.R.K., 1990. Fluoride induced changes in protein metabolism in tissues of freshwater crab, *Barytelphusa guerini*. *Environ. Poll.*, 97: 97–108.

Rajaiah, V. and Venkaiah, Y. Effect of parathion on esterase patterns of *Channa punctatus*. *J. Aqu. Biol.*, 22(1): 181–185.

Shailaja, V., Madhuri, E., Ramesh Babu, K., Rama Krishna, S. and Bhasker, M., 2008. Study of protein metabolism in hepatopancreas and muscle of prawn *Penaeus monodon* on exposure to altered pH media. *Ecology and Fisheries*, (1): 85–92.

Chapter 36

Gonadosomatic Index and Spawning Season of Snow Trout *Schizopyge esocinus* (Heckel, 1838)

☆ *Shabir Ahmad Dar, A.M. Najar, M.H. Balkhi, Mohd. Ashraf Rather and Rupam Sharma*

ABSTRACT

In the present investigation the Gonadosomatic index (GSI) and spawning season snow trout *Schizopyge esocinus* from Jhelum river, district Srinagar were studied. Observation on the variations in seasonal fluctuations in the Gonadosomatic index confirmed that the spawning period begins in May with peak in June. In the present study GSI is found to be directly proportional to spawning season and inversely proportional to post-spawning season.

Introduction

One of the methods of studying the spawning season is to follow the seasonal changes in gonad weight in relation to body weight expressed as the gonadosomatic index (Qasim, 1957; Jhigran,1961; Dadzie *et al.*, 2000; Ahirrao, 2002). Gonads undergo regular seasonal cyclical changes in weight, particularly in females. Such cyclical changes are indicative of the spawning season (Qasim, 1973; Dadzie *et al.*, 2000).

Schizopyge esocinus is commercially important. Despite its economic importance, there is paucity of information on some aspects of the reproductive biology of the species. Furthermore, *Schizopyge esocinus* is a candidate for aquaculture but due to the lack of the proper knowledge about the biology of such important fish like reproduction, feed and feeding habits etc. the culture practice of above fish is not reported from this area. The aims of these studies were to gain an understanding of the reproductive biology aspects of this species relevant to successful fishery management and aquaculture programmes. These were achieved by determining the maturity stages, Gonadosomatic index, recruitment pattern and rhythm of spawning. GSI is one of the important parameter of the fish biology which gives the detail idea regarding the fish reproduction. GSI is one such measure which can be used to assess the degree of ripeness of the gonad. GSI is the indicator of breeding period in fish (Gupta 1974) therefore, it is under taken for the study from the present area. Notable contributions on GSI are Gupta (1974, 1975), Kumthekar (1988), Reddy and Baburao (1992), Piska *et al.* (1993), Dadzie *et al.* (2000), Ahirrao (2002),Telvekar *et al.* (2005), Shankar and Kulkarni (2006), Shendge and Mane (2006, 2007) and Pawar *et al.* (2007). No attempts have no far been made to study the gonadosomatic index of *Schizopyge esocinus* from this area and this is the first kind of report on the GSI of above fish from Srinagar district.

Material and Methods

The specimen of *Schizopyge esocinus* were collected from river Jhelum lies between 33° 14° north longitude and 74°38° east (latitude) Srinagar District, Jammu and Kashmir for one year commencing from January 2007. During the present investigation 240 individuals were examined. The fishes were thoroughly washed with water and blotted completely to remove excess of water and weighed on electrical balance. The body weight of the individual fish was recorded. To study GSI the fishes were dissected for the gonads. The weight of gonads of individual fish was determined.

$$GSI = \frac{\text{Weight of the gonad}}{\text{Weight of the fish}} \times 100$$

The average gonad weight, body weight ratio were determined for each month and the values were expressed as percentages. The data are given in Table 36.1 and graphically presented in Figures 36.1 and 36.2.

Results and Discussion

Maximum GSI values were recorded around spawning season during April, May and June with peak values of 20.11 for female and 18.21 for male in May. It has been observed that *S. esocinus* is an annual breeder *i.e.* spawns once in a year. The sudden rise in GSI of from March, April (Table 36.1 and Figures 36.1 and 36.2) indicates that the fish are in the advanced maturity stage. The index rises sharply in April indicating further progress of the maturation phase. Thereafter, the GSI rises steeply during April and reaches the peak in May, suggesting the full development of gonad by that time. The GSI falls in June and there is further decline of GSI July indicates release of gonadal products during intense spawning activity. Thereafter, there were no appreciable changes in GSI from August, September and October. This suggested that the gonads were in resting phase during the period.

Table 36.1: Monthly Variation in Gonadosomatic Index (GSI) of *Schizopyge esocinus*

Month	*GSI (F)*	*GSI (M)*
January	8.78	6.89
February	10.84	8.59
March	13.28	11.36
April	15.65	13.96
May	20.14	18.21
June	10.32	9.29
July	4.82	3.56
August	0.94	0.74
September	2.09	1.68
October	3.91	2.21
November	5.44	4.59
December	7.58	6.23
Pooled mean	8.65±1.65	7.28±1.53

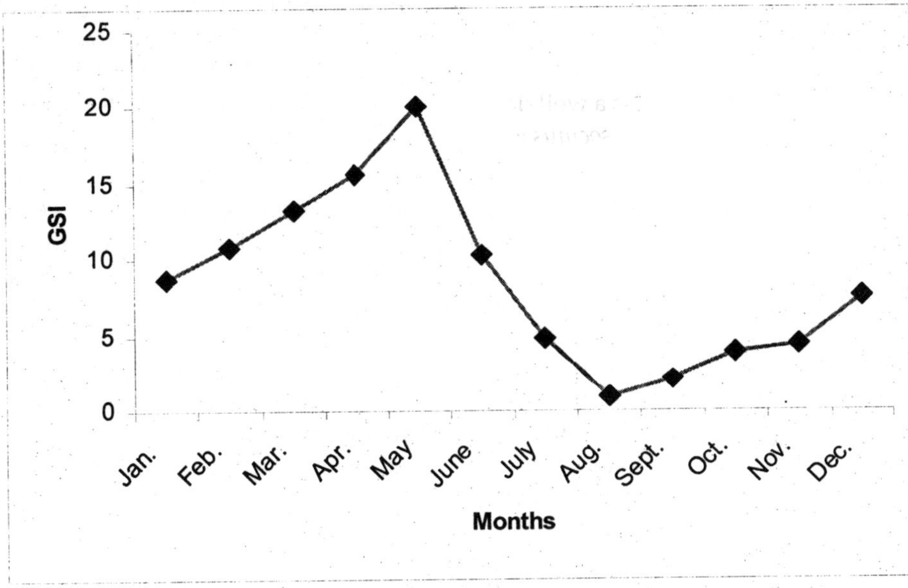

Figure 36.1: Monthly Variation in Gonadosomatic Index (GSI) of Female *Schizopyge esocinus*

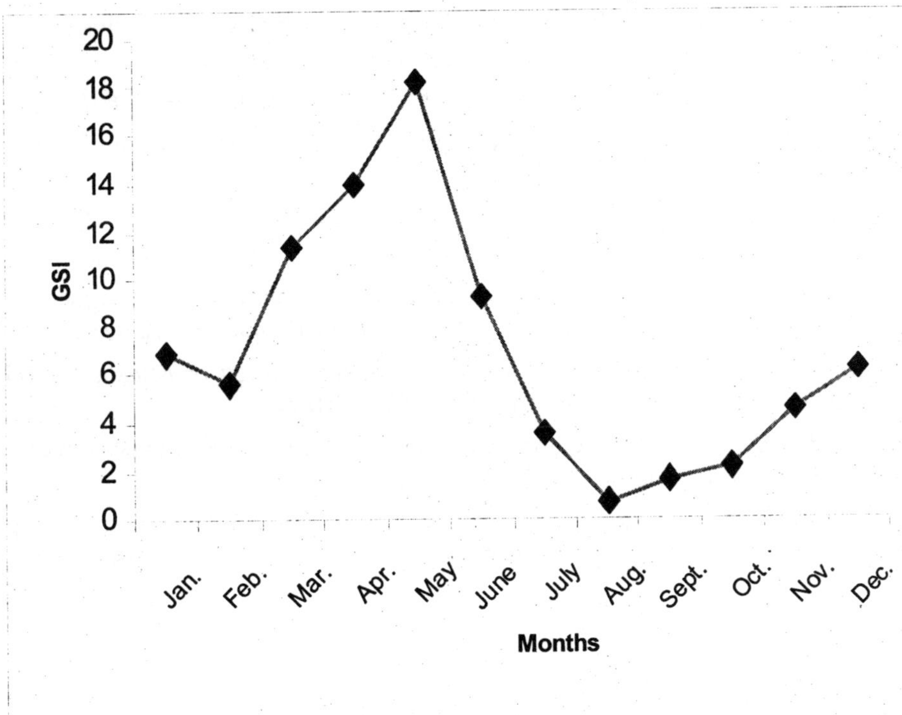

Figure 36.2: Monthly Variation in Gonadosomatic Index (GSI) of Male *Schizopyge esocinus*

The data of GSI indicate that this fish has a specific gonadal cycle. The GSI reaches its peak during spawning period. The data also indicates that *S. esocinus* has a specific season in a year, commencing in May, the spawning becomes intense during June and July. The data on GSI also indicates that *Schizopyge esocinus* has a well defined spawning season which lasts from June to July. It may also be inferred that *Schizopyge esocinus* is a total spawner. All the ripe eggs are shed within a very short period.

In present study the GSI of *Schizopyge esocinus* from river Jhelum of Srinagar showed that the weights of gonads followed regular cyclic changes which is corrected with the gamatogenesis activity in the gonads was indicated by seasonal changes in GSI. The decline following peak further suggested the onset of spawning. Similar results were obtained in *Mastacembalus armatus* (Gupta 1992), *Pampus argenteus* (Dadzie *et al.*, 2000), *Cirhinus reba* (Shendge and Mane, 2005) and *Macrones bleekeri* (Pawar and Mane, 2007). The reports are also available on many other fishes from Indian waters and abroad showing that the GSI increased with the progressive development of the gonads become ripe and the index then declined sharply in the spawning and the spent fishes (Macer, 1974; Babikar and Ibrahim, 1979; Al-Daham and Bhatti, 1979; Dadzie and Wamgilla, 1980; Jayaprakas and Nair, 1981; khan *et al.*, 1990; Veerappan *et al.*, 1997; and Ahirrao, 2002).

References

Ahirrao, S.D., 2002. Status of gonads in relation to total length (TL) and Gonadosomatic (GSI) in freshwater spiny eel (Pisces) *Mastacembalus armatus* (Lacepede) from Marathawada region, Maharashtra. *J. Aqua. Biol.*, 17(2): 55–57.

Dadzie, S., Abu-seedo, F. and Al-Shallal, T., 2000. Reproductive biology of the silver pompret, *Pampus argenteus* (Euphrasen), in Kuwait waters. *J. Appl. Icthyol.*, 16: 247–253.

Gupta, S., 1975. Some observations on the biology of *Cirrhinus reba* (Cuvier). *J. Fish. Biol.*, 7: 71–76.

Jayaprakas, V. and Nair, N.B., 1981. Maturation and spawning in the pearl spot *Etroplus surantensis* (Bloch). *Proc. Indian Nat. Sci. Acad.*, B47(6): 828–836.

Kumthekar, V.R., 1988. Biology of cyprinid fish, *Cirrhina reba* (Hamilton) from Marathwada region. *Ph.D. Thesis*, Dr. Babasaheb Ambedkar Marathwada University, Aurangabad.

Pawar, B.A., Mane, U.H. and Aher, S.K., 2007. Gonadosomatic index and spawning season of *Macrones bleekeri* (Bleeker) from Sadatpur lake, Maharashtra, India. *J. Aqua. Biol.*, 22(2) 135–138.

Ramakrishniah, M., 1992. Studies on the breeding feeding biology of *Mystus aor* (Hamilton) of Nagarjunasagar Reservoir. *Proc. Nat. Acad. Sci., India*, 63(B): 3.

Shendge, A.N. and Mane, U.H., 2005. Gonadosomatic Index and spawning season of cyprinid fish *Cirrhina reba* (Hamilton). *J. Aqua. Biol.*, 20(2): 57–59.

Veerappan, N., Ramanathan, M. and Ramaiyan, V., 1997. Maturation and spawning biology of *Amblygaster sirm* from parangipettai, Southeast coast of India. *J. Mar. Biol. Ass., India*, 39(1 and 2): 89–96.

Chapter 37

Constraints in Shrimp Farming in the North Konkan Region of Maharashtra State

☆ *A.R. Sathe, R. Pai, M.M. Shirdhankar and M.M. Gawde*

ABSTRACT

Constraints of the shrimp farmers were listed out while studying the adoption of shrimp health management practices in the North Konkan region of Maharashtra. Constraints such as high electricity charges and non-availability of credit and insurance schemes were faced by all the respondents. Price fluctuation during the time of harvest and increased cost of production were faced by 97.87 per cent of the respondents. Non-availability of shrimp seed testing laboratories in nearby areas and distant location of hatcheries as constraints were reported by 94.68 and 85.11 per cent of the respondents respectively. Disease incidence was reported by 94.68 per cent of the respondents. Most of the respondents (96.81 per cent) informed that there was decrease in shrimp rate. Inadequate government support was reported by 89.36 per cent of the respondents.

Introduction

Every business has its own constraints; like-wise shrimp farmers also have number of constraints. In order to list out various constrains faced by shrimp farmers of the North Konkan region while studying the adoption of shrimp health management practices, data on constrains were also collected during the period of four months i.e. from September to December 2007 by using the structured interview schedule from all the working shrimp farms (Sample size: 94 nos.).

Results

Constraints faced by respondents are segregated into two parts viz., constraints related to shrimp health management and other constraints as follows.

Constraints Related to Shrimp Health Management

Number of respondents facing constraints related to shrimp health management is presented in Table 37.1 and depicted in Figure 37.1. Disease incidence was the major constraint faced by 94.68 per cent of the respondents followed by distant location of hatcheries (85.11 per cent). Two-third of the respondents (65.96 per cent) did not get adequate technical know-how. Constraints such as lack of quality seed, poor water quality and silting of creeks were faced by 64.89, 37.23 and 27.66 per cent of the respondents respectively.

Table 37.1: Number of Respondents Facing Various Constraints Related to Shrimp Health Management

Sl.No.	*Constraints*	*Yes*		*No*	
		Number	*Percentage*	*Number*	*Percentage*
1.	Disease incidence	89	94.68	5	5.32
2.	Distant location of hatchery	80	85.11	14	14.89
3.	Inadequate technical know-how	62	65.96	32	34.04
4.	Lack of quality seed	61	64.89	33	35.11
5.	Poor water quality	35	37.23	59	62.77
6.	Silting of creeks	26	27.66	68	72.34

Other Constraints

Number of respondents facing other constraints is presented in Table 37.2. High electricity charges and non-availability of credit and insurance schemes were faced by all the respondents. Price fluctuation

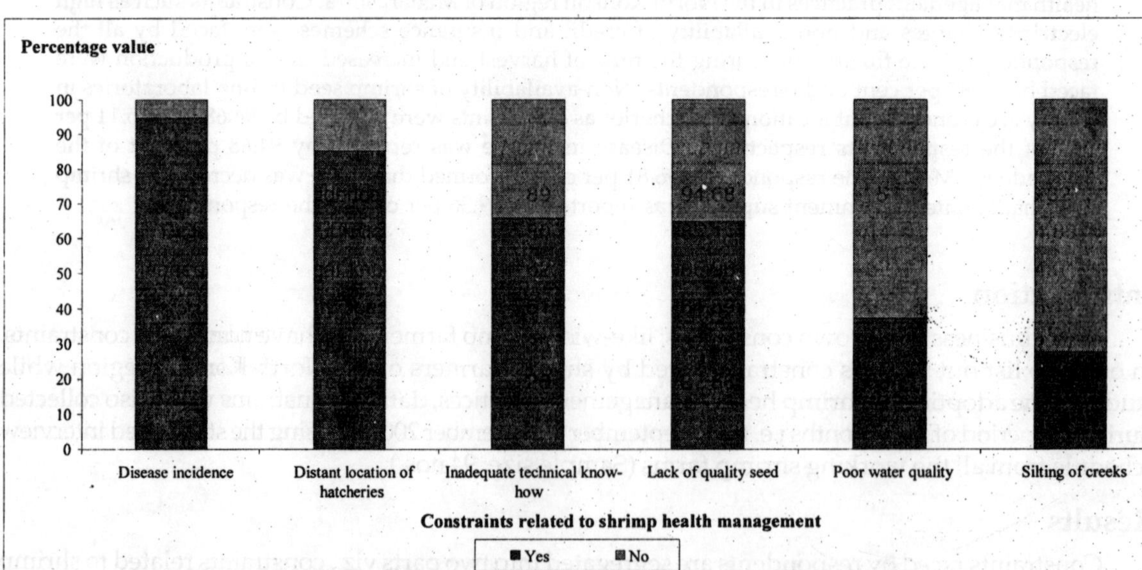

Figure 37.1: Percentage of Respondents Facing Constraints Related to Shrimp Health Management

during the time of harvest and increased cost of production were faced by 97.87 per cent of the respondents. Decreased rate of shrimp, non-availability of shrimp seed testing laboratories in nearby areas and inadequate government support were reported by 96.81, 94.68 and 89.36 per cent of the respondents respectively. Labour scarcity was reported by 46.81 per cent of the respondents.

Table 37.2: Number of respondents facing other constraints

Sl.No.	Constraints	Yes		No	
		Number	Percentage	Number	Percentage
1.	High electricity charges	94	100.00	0	0.00
2.	Non-availability of credit and insurance schemes	94	100.00	0	0.00
3.	Price fluctuation during the time of harvest	92	97.87	2	2.13
4.	Increased cost of production	92	97.87	2	2.13
5.	Decreased rate of shrimp	91	96.81	3	3.19
6.	Non-availability of shrimp seed testing lab in nearby areas	89	94.68	5	5.32
7.	Inadequate government support	84	89.36	10	10.64
8.	Labour scarcity	44	46.81	50	53.19
9.	Flood problem	38	40.43	56	59.57
10.	Poor cooperation among farmers	36	38.30	58	61.70
11.	Poaching	30	31.91	64	68.09
12.	Anti-shrimp farming lobby	28	29.79	66	70.21

Discussion

Constraints Related to Shrimp Health Management

Disease Incidence

Disease incidence was the major constraint faced by 94.68 per cent of the respondents in the present study. Kumaran and Kalaimani (2005) found disease incidence as the foremost constraint in a sample of 120 shrimp farmers in coastal districts of Tamil Nadu.

Distant Location of Hatchery

Most of the shrimp farmers (85.11 per cent) in the North Konkan Region of Maharashtra did not get the hatchery seed in nearby areas, because all the hatcheries were present out of Maharashtra State. Most of the shrimp farmers required more than six hours to transport the seed from hatchery to their farm which is not a good condition at all for the health of seed.

Inadequate Technical Know-how

In the North Konkan region of Maharashtra, though the technicians of private feed and medicine companies' were present, two-third of the respondents (65.96 per cent) did not get adequate technical know-how. Among a sample of 80 prawn farmers located in Kannamali-Chellanam and Vypin regions in Ernakulam district of Kerala state, Srinath (1996) reported that 80 per cent prawn farmers had the same constraint *i.e.* lack of technical knowledge. Ponnusamy *et al.* (2001) found that technical guidance as a problem was faced by 50 per cent of the respondents. Comparatively, less number of farmers (13.33 per cent) was observed with this constraint in the sample of 120 shrimp farmers of East Godavari district in Andhra Pradesh (Kumaran *et al.*, 2003b). Deboral *et al.* (2004) reported that 61.33 per cent of the respondents in Nagapattinam and 52 per cent of the respondents in Thanjavur pointed out the inadequacy of information as a major problem in respect of latest technologies in the sample of 150 shrimp farmers from each district. Inadequate technical know-how was ranked fifth as the constraint in the study of Kumaran and Kalaimani (2005). Thus the lack of technical know-how was the common problem faced by the shrimp farmers of the study area as that of the problems faced by shrimp farmers of other regions.

Lack of Quality Seed

Most of the farmers in the present study were dependent on middlemen for procurement of the seed, because they could not visit the hatcheries due to the economic problem. So they were not sure about the quality of seed. As a result, lack of quality seed was reported by 64.89 per cent of the respondents. In East Godavari district, the constraint of poor seed quality was faced by 67.50 per cent of the farmers (Kumaran *et al.*, 2003b). Lack of quality seed as the constraint got second rank in the study of Kumaran and Kalaimani (2005).

Poor Water Quality

Poor water quality was reported by 37.23 per cent of the respondents in the present study. Kumaran *et al.* (2003b) also reported the constraint of poor water quality with percentage value of 40.83 in East Godavari district of Andhra Pradesh.

Silting of Creeks

Silting of creeks was reported by 27.66 per cent of the respondents in the North Konkan region of Maharashtra. The same constraint got tenth rank in the study of Kumaran and Kalaimani (2005).

Other Constraints

All the respondents in the present study were suffered from the problem of high electricity charges and non-availability of credit and insurance schemes. Lack of credit and insurance as the constraint got sixth rank in the study of Kumaran and Kalaimani (2005). Lack of credit and insurance as a constraint was faced by comparatively less number (23.33 per cent) of farmers in East Godavari district (Kumaran *et al.*, 2003b) because most of the farmers obtained inputs on a credit basis from local traders through buy back arrangements. Ponnusamy *et al.* (2001) found 45 per cent respondents with the constraint of lack of credit from institutional resources. In the North Konkan region, the constraint of price fluctuation of the produce during the time of harvest was faced by 97.87 per cent of the respondents. Similarly, low price at the time of harvest and middlemen interference as constraints was faced by 61.67 per cent of the farmers of East Godavari district (Kumaran *et al.*, 2003b). Ponnusamy *et al.* (2001) found that 40 per cent of the respondents were with the constraint of high cost of seed, feed and other inputs. Similarly, increased cost of production was faced by 97.87 per cent of the respondents in the North Konkan region. In the present study, 89.36 per cent respondents reported inadequate government support. Lack of government support as the constraint got third rank in the study of Kumaran and Kalaimani (2005). In present study, poaching was reported by 31.91 per cent of the respondents. In the study of Kumaran *et al.* (2003b) poaching was reported by only five percent of the farmers. Labour scarcity was the constraint faced by 46.81 per cent of the respondents in the present study. The same constraint got ninth rank in the study of Kumaran and Kalaimani (2005). Ponnusamy *et al.* (2001) found that 55 per cent of the farmers had poor cooperation among them; similarly in the present study 38.30 per cent respondents reported this constraint. Kumaran *et al.* (2003b) also reported the constraint of poor cooperation among fellow farmers with percentage value of 48.33 in East Godavari district of Andhra Pradesh. Poor cooperation among farmers as the constraint got seventh rank in the study of Kumaran and Kalaimani (2005).

References

Deboral Vimala, D., Ramachandran, S. and Nila Rekha, P., 2004. Strengthening extension programmes in shrimp farming. *Fishery Technology*, 41: 155–158.

Kumaran, M. and Kalaimani, N., 2005. Information utilisation and extension needs assessment in shrimp farming in Tamil Nadu. Project No. CIBA/SSD/03, Technology transfer, Socio-economic aspects and informatics in brackish water aquaculture. Central Institute of Brackish water Aquaculture, Chennai, India, 27 p.

Kumaran, M., Ravichandran, P., Gupta, B.P. and Nagavel, A., 2003. Shrimp farming practices and its socio-economic consequences in East Godavari district, Andhra Pradesh, India: A case study. *Aquaculture Asia*, 8: 48–52.

Ponnusamy, K., Gopinathan, K., Kumaran, M. and Krishnan, M., 2001. Constraints in the adoption of shrimp farming in Tamil Nadu. *Applied Fisheries and Aquaculture*, 1: 103–105.

Srinath, K., 1996. Adoption of shrimp culture practices by farmers. *Seafood Export Journal*, 27: 9–12.

Chapter 38

Studies on pH and Total Dissolved Solids Fluctuations in Kaij City, Maharashtra

☆ *A.D. Chalak*

Introduction

Clean water is one of the most important needs of our bodies. It is a sad fact that something as essential to life as clean drinking water can no longer be granted to us. Most tap and well water now are not safe for drinking due to heavy industrial and environmental pollution. A systematic analysis on pH and TDS of groundwater from different places in Kaij city was undertaken. The study revealed that the water is suitable for drinking and agricultural purpose.

Beed is situated in the Deccan black basalt stone, ranges of Balaghat that constitute main range from Ahmednagar in the west, to the border of district Beed in the East.The soil of the area is rough and rocky. The district experiences semi-arid, warm and dry climate. Summers are lengthy, extending from middle of February to June. Winters are short with temperature between 12 to 20°C. Rains are inadequate and take place only during the monsoon from June to September. The annual rainfall is 666mm.

The water used for irrigation is an important factor in productivity of crop, its yield and quality of irrigated crops. The quality of irrigation water depends primarily on the presence of dissolved salts and their concentrations. Water supplies can contain dissolved organic chemical contaminants which are usually pollutants that enter water as a result of man's activities, such as insecticides, pesticides and herbicides. These are usually chronically, rather than acutely toxic to man and other species in extremely small amounts.

Contamination of drinking water has become a major challenge to the environmentalists in the developing countries. The drinking water is contaminated through the pipe distribution system or directly through groundwater. Analysis of groundwater for various parameters as per WHO standards reflect the extent of the groundwater pollution. Every year there are 1.6 million diarrhoel deaths related to unsafe water, sanitation and hygine, the vast majority among children under 5.

Materials and Methods

Water samples were collected from bore wells. The investigation was carried out for a period of thirty days. Water samples for pH and Total Dissolved Solids (TDS) were analysed with the help of Hanna's pH and TDS meter. The obtained results were compared to ISI standards (1983) as drinking water.

Results and Discussion

Water samples in general were clear, colourless and tasteless. The results obtained are depicted in Table 38.1.

Table 38.1: Fluctuations in pH and TDS in Groundwater of Kaij City in Maharashtra

Date	*pH*	*TDS (mg/l)*
01/12/2009	7.5	430
02/12/2009	8.0	420
03/12/2009	7.5	420
04/12/2009	7.3	480
05/12/2009	7.5	520
06/12/2009	7.5	580
07/12/2009	7.1	630
08/12/2009	7.3	600
09/12/2009	7.5	650
10/12/2009	7.4	620
11/12/2009	7.5	630
12/12/2009	7.6	540
13/12/2009	7.2	650
14/12/2009	7.4	560
15/12/2009	7.3	570
16/12/2009	7.5	550
17/12/2009	7.3	590
18/12/2009	7.4	570
19/12/2009	7.5	560
20/12/2009	7.3	580
21/12/2009	7.3	560
22/12/2009	7.5	540
23/12/2009	7.1	580
24/12/2009	7.2	490
25/12/2009	7.8	460
26/12/2009	7.7	450
27/12/2009	7.4	430
28/12/2009	7.1	500
29/12/2009	7.5	530
30/12/2009	7.3	520

The pH of groundwater samples show mild alkaline range of 7.1 to 8.0 indicating the presence of basic salts. The pH was in the range of suitable drinking water standards with minimum 7.1 and maximum 8.0. In general water with pH < 7 is considered acidic and with pH > 7 is considered basic. The normal range of pH for groundwater system is 6 to 8.5.

In general the water with low pH (< 6.5) could be acidic, soft and corrosive. A water with pH (> 8.5) indicates that the water is hard.

Most groundwater contain dissolved carbon dioxide, bicarbonate and carbonate and have pH values between 5 and 9 (Boyd and Tucker, 2009). In general, groundwater in contact with silicate minerals have low concentration of bicarbonate, a relatively high carbon dioxide content and consequently a lower pH than water from carbonate rock deposits. High pH values (> 8.5) are usually associated with groundwater that have a high sodium carbonate content. The pH increases with active decomposition of organic matter (Khanna and Butiani, 2007).

Many workers such as Praharaj *et.al.*(2004), Tripathi,J.K. (2003), Sambasiva Rao (1997), Mariappan *et al.* (2005), Suryawanshi *et al.* (2004), Lokhande *et al.* (2009) have been carried exhaustive study on groundwater quality. Lokhande *et al.* (2009) observed pH of groundwater of Latur city in the range 7.0 to 8.0 and TDS in the range 170 to 1820mg/l. The pH values corroborate with the present study. The TDS value of the sample water of Latur city is above permissible limit. The ISI standard for TDS is 500mg/l and the maximum permissible limit is 1500mg/l (WHO, 1994).

From the above observations, it is concluded that pH and TDS were in the permissible limit as recommended by ISI (1983). Hence the water is suitable for drinking and agricultural use.

References

Boyd, C.E. and Tuckor, C.S., 2009. *Pond Aquaculture Water Quality Management*. Springer International Edition, 700 pp.

ISI, 1983. *Indian Standards Specification for Drinking Water*. ISI 10500.

Khanna, D.R. and Bhutiani, R., 2007. *Limnological Modelling: A Case Study of River Suswa.* Daya Publishing House, Delhi, pp. 301.

Lokhande, M.V., Dande, K.G., Karadkhele, S.V., Rathod, D.S. and Shembekar, V.S., 2008. Studies on groundwater quality of Latur city, Maharashtra. *Ecology and Fisheries,* 1(1): 97–100.

Mariaappan, P.V., Yegnaraman, V. and Vasudevan, T., 2000. Groundwater fluctuation within water table level in Thiruppathur block of Sivagangi district (T.N.). *Poll. Res.,* 19(2): 225–229.

Sakhare, V.B., 2007. Studies on groundwater quality in Tuljapur city, Maharashtra. *Geobios,* 34(2–3): 205–206.

Sambasiva Rao, T., 1997. Sustainability of groundwater for irrigation in Chandragiri block, Distt. (A.P.). *Ecology,* 11(10): 1–9.

Suryawanshi, M., Bhagwan, Kalyankar, K.B. and Pande, B.N., 2004. Groundwater analysis in an industrial zone Chikalthana (Aurangabad). *Poll. Res.,* 23(4): 649–653.

Tripathi, J.K., 2003. Groundwater histochemistry in and around Bhanjabjihar, Ganjam district (Orissa). *Poll. Res.,* 22(2): 185–188.

Trivedy, R.K. and Goel, P.K., 1986. *Chemical and Biological Methods for Water Pollution Studies.* Environmental Publications, Karad, Maharashtra.

Chapter 39

Effect of *Piper longum* Extract on the Gross Primary Productivity of an Angel Fish Mass Culture System

☆ *K. Kannan and G. Rajasekaran*

ABSTRACT

An aqueous herbal extract of long pepper (*Piper longum*) was used to control argulosis and Lernaeosis in an Angel fish (*Pterophyllum* sp.) mass culture system. The gross primary productivity analysed in the experimental tanks during a 24 hour cycle at an interval of three hours for thirty days did not differ from that of the control tank indicating that the extract of long pepper is a potential curative agent of argulosis and Lernaeosis without having an adverse effect on the gross primary productivity of the culture system.

Introduction

Aquaria have captured the imagination of man for a long time. The Romans were the first to have setup an aquarium around the year 50 (Brunner, 2003). After this time, interest in aquarium fish keeping has grown tremendously and is now a popular hobby around the world. In the United States aquarium keeping is the second most popular hobby after stamp collecting (Riehl and Baensch, 2004). Developing countries like India are fast catching up in aquarium keeping as well as in promoting this trade.

Substantial economic loss has been experienced in aquaculture systems mainly due to disease outbreak caused by pathogens and parasites (Chakraborthy and Chattopadhyay, 1998). These diseases are controlled at various levels, which include identification with diagnosis, treatment with therapeutants, prevention with immuno modulators and husbandry factors (Brackett and Little, 1995).

Conventionally chemotherapeutics are employed for disease control (Subhashinghe and Sheriff, 1995). Some of the common antibiotics and sterilizing agents are oxytetracycline, erythromycin, amoxycillin, quinolones, pyrethrins, hydrogen peroxides, organophospates, formalin, malachite green, chloromine, potassium permanganate and others (Brackett and Little, 1995). There are many disadvantages when fish diseases are cured by chemotherapeutants as the pathogens can develop resistance to antibiotics very quickly (Kong, 1995). Since chemotherapy enforces many ill-effects on the fishes as well as on the culture system, alternate effective and environmental friendly herbal medicines are now being chosen to treat diseases. Medicinal plants have been widely in use for a long time in India to treat various disorders including infectious diseases like epizootic ulcerative syndrome, myxobiosis, gyrodactylosis and argulosis (Ocampo and Jiminez, 1993; Dey and Chandra, 1994). Rath (1990) has listed the efficacy of 15 herbal medicines in the cure of fish diseases.

The use of chemical therapeutants or herbal extracts may cause an imbalance in the gross primary productivity of a mass culture system. Therefore this study is aimed at investigating the same in an Angel fish mass culture system to control argulosis and Lernaeosis using the herbal extract of Long Pepper.

Materials and Methods

The experiments were conducted for thirty days in six brick-walled, clay bottomed cement tanks with water recirculation facilities. The tanks had an average measurement of 2.50 × 2.5 × 1m with an average capacity of 8228 liters each.

The gross primary productivity was estimated in both the control and experimental tanks. The experimental tanks were used for stocking an Angel fish measuring 15 to 20mm total length at a stocking density of 4 fishes/litre. One isolated tank was used as control whereas the other five were experimental tanks with water recirculation among themselves.

The experiment to determine the optimum concentration of herbal extract of Long Pepper was conducted in five circular glass troughs; the volume of each trough was 10.13 liters. Each of them was filled with 10 liters of water collected from the Angel fish experimental tanks. Five an Angel fishes were introduced into all the tanks. The No.1 glass trough was maintained as control and other troughs (Nos. 2-5) as experimental. To these experimental glass troughs, 0.01 per cent, 0.02 per cent, 0.03 per cent, and 0.04 per cent of the herbal extracts were added respectively. The concentration of 0.03 per cent herbal extract was determined to be optimal and the same concentration was maintained in the experimental tanks.

The primary productivity estimation was based on the method described by Michael (1984). A one metre length PVC pipe of 25mm diameter with two 'T' joints was used. The PVC pipe was kept vertically in the experimental tank. The first 'T' joint was fixed at a length of 0.15m from the top of the pipe and the second 'T' joint was fixed at a length of 0.65m from the first 'T' joint. These joints were used to indicate the surface and bottom levels and also to suspend the 'light' and 'dark' bottles.

Results and Discussion

Atmospheric Temperature

Atmospheric temperature measured from 06.00 hrs to 18.00 hrs during the study period varied from 30°C to 36°C (Table 39.1).

Water Temperature

The water temperature at the surface and bottom of the control and experimental tanks from 06.00

hrs to 18.00 hrs are represented in Table 39.1. The water temperature was the same at both the surface and bottom of the Angel fish culture tanks. This could be due to the shallow depth of 1m and also due to water recirculation. The temperature varied from 29°C to 30°C.

Table 39.1: Average Atmospheric and Water Temperatures

Sl.No.	*Time (Hrs)*	*Average Atmospheric Temperature (°C)*	*Average Water Temperature (°C)*
1.	06–09	30.8	29.5
2.	09–12	31.5	29.9
3.	12–15	33.2	30.8
4.	15–18	33.4	31.2

Gross Primary Productivity in Control Tank

The Gross Primary Productivity in control tank is represented in Table 39.2. The gross primary productivity from 06.00 hrs to 09.00 hrs at the surface was either 0.28 mgc/m^3/3hrs or 0.56 mgc/m^3/3hrs (mean: 0.46 mgc/m^3/3hrs), whereas at the bottom, it was either 0.28 mgc/m^3/3hrs or nil (mean: 0.18). From 09.00 hrs to 12.00 hrs, the gross primary productivity at the surface was either 0.56 mgc/m^3/3hrs or 0.84 mgc/m^3/3hrs (mean: 0.68 mgc/m^3/3hrs), while at the bottom it ranged from 0.56 mgc/m^3/3hrs (mean: 0.22 mgc/m^3/3hrs). From 12.00 hrs to 15.00 hrs the minimum value of gross primary productivity at the surface was 0.56 mgc/m^3/3hrs and the maximum was 1.4 mgc/m^3/3hrs (mean: 0.83 mgc/m^3/3hrs). At the bottom it fluctuated between 0 and 0.56 mgc/m^3/3hrs (mean: 0.26). Between 15.00 hrs to 18.00 hrs at the surface, the lowest value of the gross primary productivity was 0.56 mgc/m^3/3hrs and the highest was 1.12 mgc/m^3/3hrs (mean: 0.73 mgc/m^3/3hrs), whereas at the bottom it fluctuated from 0 to 0.56 mgc/m^3/3hrs (mean: 0.38 mgc/m^3/3hrs).

Table 39.2: Gross Primary Productivity in Control and Experimental Tanks

Sl.No.	*Time (Hrs)*	*Primary Productivity (mgc/m³/3hrs)*			
		Control Tank		*Experimental Tank*	
		Surface	*Bottom*	*Surface*	*Bottom*
1.	06–09	0.46	0.18	0.46	0.19
2.	09–12	0.68	0.22	0.67	0.20
3.	12–15	0.83	0.26	0.82	0.22
4.	15–18	0.73	0.38	0.69	0.19

Gross Primary Productivity in Experimental Tanks

The Gross Primary Productivity in experimental tank is given in Table 39.2. The gross primary productivity values at the surface of the experimental tanks from 06.00 hrs to 09.00 hrs was either 0.28 mgc/m^3/3hrs or 0.56 mgc/m^3/3hrs, (mean: 0.46 mgc/m^3/3hrs), whereas at the bottom, it was either 0.28 mgc/m^3/3hrs or nil (mean: 0.19 mgc/m^3/3hrs). The gross primary productivity at the surface was either between 09.00 hrs to 12.00 hrs was either 0.56 mgc/m^3/3hrs or 0.84 mgc/m^3/3hrs, (mean: 0.67 mgc/m^3/3hrs). At the bottom it varied from 0 to 0.56 mgc/m^3/3hrs (mean: 0.20 mgc/m^3/3hrs). Between 12.00 hrs to 15.00 hrs at the surface, the minimum value of gross primary productivity was

0.56 mgc/m^3/3hrs and the maximum was 1.4 mgc/m^3/3hrs (mean: 0.82 mgc/m^3/3hrs). At the bottom, the lowest and highest values of the gross primary productivity were 0 and 0.56 mgc/m^3/3hrs respectively (mean: 0.22 mgc/m^3/3hrs). At the surface, the minimum value of gross primary productivity during 15.00 hrs to 18.00 hrs was 0.56 mgc/m^3/3hrs and the maximum was 1.12 mgc/m^3/3hrs (mean: 0.69 mgc/m^3/3hrs) whereas, at the bottom it ranged from 0 to 0.56 mgc/m^3/3hrs with a mean of 0.19mgc/m^3/3hrs.

Based on the results, it is evident that in general the gross primary productivity at the surface and bottom during 06.00 and 18.00 hrs in both the control and experimental tanks was the same with a narrow range of variation.

At the surface, the gross primary productivity was relatively higher in both the control and experimental tanks due to greater intensity of sunlight. At the same time the gross primary productivity at the bottom was relatively reduced due to low intensity of sunlight (Jhingran and Ahmad, 1998). In both the control and experimental tanks the gross primary productivity was found to be higher at the surface than at the bottom during 06.00–09.00 hrs. This difference may be due to higher intensities of sunlight at the surface than at the bottom.

Between 09.00 hr and 12.00 hrs in the control as well as in the experimental tanks at the surface and bottom, the gross primary productivity increased remarkably as compared to 06.00–09.00 hrs. However, at the bottom, the gross primary productivity decreased due to low intensity of the sunlight (Odum, 1971 and Jhingran, 2002). During 12.00–15.00 hrs in both the control and experimental tanks, at the surface and bottom, the gross primary productivity was relatively more as compared to that of 09.00–12.00 hrs. This is perhaps due to the maximum intensity of the sunlight. In the control as well as in experimental tanks between 15.00 and 18.00 hrs at both the surface and bottom, the gross primary productivity was relatively less as compared to that of 12.00–15.00 hrs. However, at the bottom, the gross primary productivity decreased due to low intensity of the sunlight. So, at the bottom of the control and experimental tanks the gross primary productivity was less than that of the surface.

Conclusion

Gross primary productivity studies of the angel fish mass culture system point to the fact that the impact of herbal extract of Long Pepper in the experimental tanks on the primary productivity in general was not different from the control tank. Further, the gross primary productivity analysed during 24 hrs cycle at an interval of 3 hrs also did not differ from the control tank. This shows that extract of Long Pepper which is a curative agent of argulosis and lernaeosis will not have any adverse effect on the gross primary productivity of the culture system. Hence, it is recommended that Long Pepper extract which is a potential curative agent for eradicating argulosis and lernaeosis can be effectively used in the mass production culture techniques of Angel fish.

References

Brackkett, J.B. and Little, J.M., 1995. Recent trends in fish chemotherapeutants. In: *Proceedings of the Workshop on Aquaculture for 2000 A.D.*, Singapore, pp. 225–227.

Brunner, B., 2003. *The Ocean at Home*. Princeton Architectural Press, NewYork, pp. 21–22.

Chakraborthy, C. and Chattopadhyay, A.K.R., 1998. Turmeric and neem leaf extract in the management of bacterial infection in African cat fish, *Clarias gariepinus*. *Fishing Chimes*, 18: 38–42.

Dey, R.K. and Chandra, S., 1994. New trends in fish diseases management through application of herbal materials. *Fishing Chimes*, 14: 13–14.

Jhingran, V.G., 2002. *Fish and Fisheries of India*. Hindustan Publishing Corporation, New Delhi, pp. 728.

Jhingran, V.G. and Ahmad, S.H., 1998. Aquatic productivity related with fish yields from water. In: *Aquaculture Productivity*, (Eds.) V.R.P. Sinha and H.C. Srivastava. Oxford and IBH Publishing, New Delhi, pp. 285–294.

Kong, C.S., 1995. Immune enhancers in control of disease in aquaculture. In: *Proceedings of the Workshop on Aquaculture for 2000 A.D.*, Singapore, pp. 229–233.

Michael, P., 1984. *Ecological Methods for Field and Laboratory Investigations*. Tata McGraw-Hill Publishing Co. Ltd., New Delhi, pp. 404.

Ocampo, A.A. and Jiminez, E.M., 1993. Herbal medicine in the treatment of fish diseases in Mexico. *Veterinaria Mexico*, 24: 291–295.

Odum, E.P., 1971. *Fundamentals of Ecology*, 3rd Edn. W.B. Saunders Co., Philadelphia, pp. 574.

Rath, K.K., 1990. Prevention and control of fish disease by herbal medicines. *Fishing Chimes*, 9: 38–39.

Riehl, R. and Baensch, H.A., 2004. *Aquarium Atlas*. Tetra Press, Germany, pp. 892.

Subasinghe, R.P. and Sheriff, M., 1995. Impact of disease on aquaculture. In: *Proceedings of the Workshop on Aquaculture for 2000 A.D.*, Singapore, pp. 56–59.

Chapter 40

A Study on the Development of Immunity in the Common Carp, *Cyprinus carpio* by Neem Azal Formulation

☆ *N. Muthumurugan, M. Pavaraj and V. Balasubramanian*

ABSTRACT

Aquaculture and agriculture are not strictly parallel developments in food production, eventhough food gathering, hunting, and fishing might have started at about the same time in human history. The complexity of aquaculture as a multi disciplinary activity, even more complex than agriculture is perhaps one of the reasons for the late start of modern aquaculture. Fisheries can be taken as a major economic activity in the country as it supports nutrition, employment and foreign exchange. The present investigation was conducted to study the immunoprotective effect of neem azal formulation on common carp, *Cyprinus carpio* against heat-killed *Aeromonas hydrophila,* as an antigen. 43 per cent more survival rate was observed in neem azal treated fishes than the control.The neem azal (10 ppm) treated fish favoured the production of antibodies and relatively high protection than the control fish group.

Introduction

Aquaculture has been expanding rapidly in many developing countries in an attempt to increase fish production and earn foreign exchange (Chythanya *et al.,* 1999). Large scale mortalities of fish often occur in ponds and lakes due to environmental pollution or stress followed by microbial infection (Michael, 2002). The common carp, *Cyprinus carpio* is commercially important and widely cultured species, because of the high tolerance to environmental fluctuations (Jhingran and Pullin, 1988). One of the major constraints for the development of aquaculture is the loss due to disease. Bacteria and viral diseases often cause serious mortalities (Karunasagar and Karunasagar, 2001). During an outbreak of disease among Indian major carps that had more than 75 per cent morality, the causative agent was found to be *Aeromonas hydrophila* (Katoch *et al.*, 2003). Neem products are presently used extensively by rural farmers as indigenous technology for controlling a variety of pathogens in fish culture systems (Das *et al.*, 2002).Therefore, the present study deals with the immunoprotective effect of neem azal formulation in *C. carpio* by estimating the survival rate, antibody titre and relative level of protection.

Materials and Methods

Experimental Animal and Maintenance

Fingerlings of common carp, *C.carpio*, (10±1g) have been collected from Meenakshi fish farm, Madurai, Tamil Nadu (India). The collected fish were transported to the laboratory in polytene bags containing aerated water. They were acclimated to the laboratory conditions in glass aquaria for 15 days. The fingerlings of common carp were fed with artificial feed, twice a day.

Experiment

Neemazal (the neem formulation) a brown, viscous liquid containing 1 per cent Azadirachtin was purchased from local market and prepared into 10ppm concentration as suggested by earlier workers. 100µl of the neem azal was injected into the experimental fish. Two days later, 0.1ml of 10^6 CFU/ml of heat killed *Aeromonas hydrophila* was injected into the fish intraperitoneally. The control groups received only heat killed *A. hydrophila* without neem formulation.

The number of fish survived to the total number of fish, expressed in percentage can be calculated as the survival rate. The antibody titre was recorded and expressed as $\log_2$ antibody titre of serum following bacterial agglutination assay (Michael, 2000). The same volume of fresh culture of *A. hydrophila* was injected into the fish to observe relative level of protection (RLP), following the formula:

$$RLP = 1 - \frac{\text{Total no. of mortality of fish in the experiment}}{\text{Total no. of mortality of fish in the control}} \times 100$$

Results and Discussion

The cumulative survival and mortality rate of the fingerlings of the common carp, *C. carpio* treated with neem azal was shown in Table 40.1. The control fish group caused 29 per cent mortality on 5th day and 57 per cent mortality (43 per cent survival) on 10th day after inoculation. The neem Azal treated fish exhibited 14 per cent mortality on 10th day. The neem formulation treated groups showed 75 per cent less mortality than the control untreated group ($P<0.05$, significant). Similar observations of significant difference in the control and experimental groups were reported (Saravanakumar *et al.*, 2007). Similar results were also obtained by various workers. The fingerlings injected with 100µl of *A. hydrophila* revealed 80 per cent survival rate in control group and 100 per cent survival in experimental groups fed with vitamin C (Narmadha *et al.*, 2007). Vicco turmeric vanishing cream treated fish showed 100 per cent survival in experimental fishes (Vadivelmurugan *et al.*, 2007).

The control group showed maximum antibody response (0.602±0.301) on 10th day and the neem formulation treated groups showed the peak antibody titre (1.304±0.174) on 15th day (Table 40.2). Therefore, maximum antibody production was observed in the experimental groups than the control.

Similar observations were observed by Kaliraj and Balasubramanian, (2006) in *C. carpio* treated with *Albizia lebbeck* and *Gymnema sylvestre.* Saravanakumar *et al.* (2007) stated that the plant extract administered experimental groups showed the peak antibody response on 21st day.

Table 40.1: The Cumulative Percentage Mortality Rate of the Fingerlings of *C. carpio* IP Administered with 0.1 ml of 10^6 CFU/ml of Heat Killed *A. hydrophila* 100µl of 10ppm Neem Azal Formulation was Administered 2 Days Prior to the Injection of Heat Killed *A. hydrophila*

Name of the Neem Formulation	*Dose (ppm)*	*Days after Administration*						*Total Mortality Rate (%)*
		0	*5*	*10*	*15*	*20*	*25*	
Control	0	0	29	57	57	57	57	57
Neem azal	10	0	0	14	14	14	14	14

Table 40.2: Antibody Titre (Log_2 values) of *C. carpio* IP Administered with 0.1ml of 10^6 CFU/ml of Heat Killed *A. hydrophila.* 100µl of 10ppm neem azal formulation was administered 2 days prior to the injection of heat killed *A. hydrophila*

Name of the Neem Formulation	*Dose (ppm)*	*Days after Administration*					
		0	*5*	*10*	*15*	*20*	*25*
Control	0	0	0.401±0.174	0.602±0.301	0.401±0.174	0.401±0.174	0.301±0
Neem azal	10	0	0.602±0.301	0.401±0.174	1.304±0.174	0.602±0.301	0.903±0.301

The control group showed less protection and the Neem Azal treated groups showed 14.3 per cent protection. The neem formulation entered into the fish and induce the immune cells. Lipton, (1997) reported the immunostimulants are capable of preventing attachment of bacteria to the gut. In addition, they make bacteria to clump together, which are easily expelled from the digestive system. The enzymes are capable of destroying the bacterial cell walls and membranes selectively without damaging host cells. So, the plant extracts treated groups, did not be affected by the bacteria and showed 100 per cent protection. The neem formulations are potential for injection to fish. However, appropriate field trials remain necessary prior to using neem formulations in fish farms.

Table 40.3: Relative Level Protection of Neem Azal Formulation Injected 2 Days Prior to the Administered Against 0.1ml of 10^6 CFU/ml Fresh Culture of *A. hydrophila.* The observations were made 25 days after inoculation of *A. hydrophila.*

Name of the Neem Formulation	*Dose (ppm)*	*No. of Fish Died*	*Relative Level of Protection (%)*
Control	0	7	0
Neem azal	10	6	14.3

Acknowledgement

The authors express their heartfelt thanks to the Management, the Principal and Head of the Department of Zoology for providing necessary facilities to carryout this work.

References

Chythanya, R., Nayak, D.K. and Venugopal, M.V., 1999. Antibiotic resistance in aquaculture. *INOFISH International*, 6: 31–33.

Jhingran, V.G. and Pullin, R.S.V., 1988. *A Hatchery Manual for the Common Chinese and Indian Major Carps*. ADB/ICLARM, Manila, Phillipines.

Karunasagar, I. and Karunasagar, I., 2001. *Sustainable Indian Fisheries*, (Ed.) T.J. Pandian, p. 166–168.

Katoch, R.C., Sharma, M., Pathania, D., Verma S., Chahota, R and Mahajan, A., 2003. Recovery of bacterial and mycotic fish pathogens from carps and other fish in Himachal Pradesh. *Indian J. Microbiol.*, 43(1): 65–66.

Kaliraj, P. and Balasubramanian, V., 2006. Immuno protective effect of a few medicinal plant extract on disease induced common carp, *Cyprinus carpio* (L). *M.Phil. Dissertation*, A.N.J.A College (Autonomous), Sivakasi.

Lipton A.P., 1997. Disease management using immunostimulants and other additives CMFRI, Vizhinjam.

Michael, R.D., 2000. *State Level Workshop on Immunological Techniques Lab Procedures*. Post Graduate and Research Department of Zoology and Microbiology, The American College, Madurai, 4: 9–10.

Narmadha, V., Anand, M. and Balasubramanian, V., 2007. Development of immunity by ascorbic acid in common carp, *Cyprinus carpio*. In: *Proceedings of the National Seminar on Applied Zoology*, Feb. 7–9: 142–144.

Saravanakumar, C., Muthumurugan, N. and Balasubramanian, V., 2007. Development of immunity by chosen medicinal plants on the fingerlings of common carp, *Cyprinus carpio* (L). In: *Proceedings of the National Seminar on Applied Zoology*, Feb. 7–9: 232–236.

Vadivelmurugan, S., Pavaraj, M. and Balasubramanian, V., 2007. Therapeautic effect of vicco turmeric vanishing cream on ulcer induced common carp, *Cyprinus carpio*. In: *Proceedings of the National Seminar on Applied Zoology*, Feb. 7–9: 133–136.

Chapter 41

Toxic Effect of Imidacloprid on the Lipid Metabolism of the Estuarine Clam, *Katelysia opima* (Gmelin)

☆ *V.B. Suvare, A.S. Kulkarni and M.V. Tendulkar*

ABSTRACT

The indiscriminate use of insecticides has caused serious pollution problems of aquatic ecosystems. Imidacloprid is found in a variety of commercial insecticides like Admire, Confidor, Premier etc., and mainly used to control sucking insects such as rice hoppers, aphids, thrips, white flies, termites and some species of beetles. Further, it is known to cause apathy, myatonia, tremor and myospasms in humans. Toxic effects of Imidacloprid were estimated by selecting *Katelysia opima* as an animal model. Effect of Imidacloprid on the total lipid content of gill, mantle, hepatopancreas, foot, male gonad and female gonad of estuarine clam, *Katelysia opima* was studied. The clams were exposed to 86.6 ppm Imidacloprid for acute treatment. It was found that, there was increase in lipid content in gill in LC_0 group as compared to the control group and decrease in lipid content in gill, mantle, male gonad, female gonad of LC_0 and LC_{50} group than the control group. The lipid content was increased in hepatopancreas and foot in LC_{50} group compared with LC_0 group.

Introduction

Imidacloprid ($C_9H_{10}ClN_5O_2$) is a systemic, chloronicotinyl insecticide used to control the sucking insects like rice hoppers, aphids, thrips, white flies, termites, turf insects, soil insects and some beetles. It is most commonly used on rice, cereals, maize, potatoes, vegetables, sugar beets, fruits, cotton, hopes and turf. It acts on the nervous system of insect which causes a blockage in the neural pathway. This blockage leads to the accumulation of acetyl choline, an important neurotransmitter, resulting in the insect's paralysis and eventually death. It is effective on contact and via stomach action (Kidd and

James, 1994). It is found in the variety of commercial insecticides (Meister, 1995). In market, it is available in the form of the products like Admire, Confidor, Premier, Provado, etc., with Imidacloprid as an active ingredient. In Ratnagiri, it is mainly used against mango hoppers, mango mealy bugs, aphids and other insect pests of mango. Due to this, Imidacloprid is getting concentrated in the aquatic bodies.

It is bioaccumulated in estuarine fauna like clams. Clams are abundant along the coast of Ratnagiri (Maharashtra) and are important with reference to the food value. Shells are mainly used as a raw material for lime factories along the coast. In spite of the fact, very less attention has been paid by researchers with regard to effect of pollution on estuarine clams. Clams are known to accumulate pesticides without getting killed easily and have relatively long life span. Since biochemical assessment is a useful tool to measure environmental quality, the present work is aimed to study the effect of Imidacloprid on protein metabolism in estuarine clam, *Katelysia opima* (Gmelin).

Various biochemical changes, under stress conditions have a great significance, since they catalyze and control the formation of various intermediates, which are indispensable to all the normal physiological processes. Rao and Rao (1981), reported decrease in the lipid levels in muscles, gill, liver and brain of the fish *S. mossambicus* exposed to methyl parathion. Muley (1991), observed significant decrease in lipid content of soft tissue like foot, gonad and mantle of freshwater clam, *Indonaia caerulens* under Endosulfan stress. Rao and Ramaneswari (2000) reported decrease in lipid content in tissue of *Labeo rohita* after exposure to Endosulfan and Monocrotophos. Monoharan and Subbiah (1982) observed decrease in lipid content of Barbus stigma exposed to Endosulfan. In the same fish, Ghosh (1986) reported 33.64 per cent decrease in lipid content due to Nuvan exposure. Muley and Mane (1989) observed decrease in lipid content in posterior adapter mussels, hepatopancreas and gills in freshwater Lamellibranch mollusc, *Lamellidens marginalis* after an acute exposure to Endosulfan. Mane and Gokhale (2002) observed significant alterations in lipid content of *M. meretrix* and *K. opima* after exposure to fluoride.

Materials and Methods

The experimental clams (*Katelysia opima*) used for the present study were collected from Bhatye estuarine region, Ratnagiri coast, Maharashtra state. Clams of medium size (4.0–4.4cms.) were selected and brought to the laboratory and stocked in the plastic containers containing filtered, aerated estuarine water for 48hrs. Well acclimatized clams to the laboratory condition were grouped in ten and kept into plastic containers containing five liters filtered estuarine water. Static bioassay tests were conducted for 96hr. by using Imidacloprid (17.8 per cent SL). For every experiment, a control group of clams was also run simultaneously.

The toxicity tests were repeated for three times and LC_0 and LC_{50} values were determined. After acute exposure to Imidacloprid the various tissues (gill, mantle, hepatopancreas, foot, male and female gonad) of live clams were pooled, weighed and dried in an oven at 70°C until a constant weight was obtained. Oven dried tissues were used for biochemical analysis. Live clams were analyzed for total lipids. (Barnes and Black stock, 1973.)

Results

Imidacloprid induced alterations in the total lipid content in the different organs of estuarine clam, *Katelysia opima* are showed in Table 41.1. Control group showed lipid content in ascending order of, hepatopancreas < female gonad < male gonad < gill < mantle < foot containing 7.556±0.1501, 7.918±0.125579, 8.706±0.132117, 8.882±0.10198, 9.03±0.178885, 9.832±0.127554mg of lipid/100mg dry wt. respectively.

Table 41.1: Imidacloprid Induced Alterations in the Total Lipid Content of *K. opima* After Acute Exposure

Tissue	*Control*	*LC_0 Group*	*LC_{50} Group*
Gill	8.82±0.10198	9.888±0.085849 (12.10)**	6.618±0.143771 (–24.96)*
Mantle	9.03±0.178885	7.298±0.12498 (–19.18)*	5.372±0.216264 (–40.50)*
H.P.	7.556±0.1501	6.470±0.182346 (–14.37)*	7.172±0.130843 (–5.08)*
Foot	9.832±0.127554	7.684±0.207437 (–21.84)*	8.30±0.225499 (–15.58)*
M.Gonad	8.706±0.132117	6.716±0.105024 (–22.85)*	5.324±0.216171 (–20.72)*
F.Gonad	7.918±0.125579	6.232±0.156109 (–21.29)*	4.596±0.212908 (–41.95)*

* Values in parenthesis are percentage difference S.D. of five readings.

In LC_0 (38.5ppm) group, lipid content was present in ascending order of, female gonad < hepatopancreas < male gonad < mantle < foot < gill with 6.232±0.156109, 6.470±0.182346, 6.716±0.105024, 7.298±0.12498, 7.684±0.207437, 9.888±0.085849mg of lipid/100 mg dry wt.

In LC_0 group, as compared to control there was 12.12 per cent increase in lipid content in gill. 23.23 per cent, 21.84 per cent, 21.29 per cent, 19.18 per cent and 14.37 per cent decrease in lipid content in male gonad, foot, female gonad, mantle and hepatopancreas respectively.

In LC_{50} group, lipid content was present in ascending order of, female gonad < male gonad < mantle < gill < hepatopancreas < foot, containing 4.596±0.21290, 5.324±0.216171, 5.372±0.216264, 6.618±0.143771, 7.172±0.130843, and 8.30±0.225499mg of lipid/100mg dry wt. respectively. As compared to control, there was 41.95 per cent, 40.50 per cent, 39.14 per cent, 24.88 per cent, 15.58 per cent, and 5.08 per cent decrease in lipid content in female gonad, mantle, male gonad, gill, foot and hepatopancreas respectively.

Discussion

In present study, the clams from LC_0 grouped showed significant decrease in lipid content in mantle, hepatopancreas, foot, male gonad and female gonad but increase in gill. While in LC_{50} group it was decreased in gill, mantle, hepatopancreas, foot, male gonad and female gonad. The lipid level decreased may be due to an increased lipolysis to meet higher energy demand to overcome toxicant stress. The decrease in lipid content might be due to the utilization of lipid to meet the additional energy requirement under pesticide stress. (Rao and Rao, 1981; Rao *et al.*, 1985). Saradamani and Kamalaveni (2002) reported that reduction in lipids reveal their utilization for energy yielding processes to overcome pesticide stress.

In the present study gill showed increase in lipid level in LC_0 group as compared to control group. Increase in lipid content may be due to shift in metabolism towards lipid content through acetyl–Co-A. This increasement may be due to increased lipogenesis through formation of peroxides and accumulation of pesticides in lipids (Bhaktavathsalam and Reddy, 1983). Swami *et al.* (1983) suggested that, the flux of carbon through the Kreb's cycle can be controlled by the relative channeling of Glucose-6 phosphate through pentose phosphate pathway and generated acetyl co-A into the lipid biosynthesis pathway. The two metabolic pathways are complementary to each other as one provides NADPH required for lipogenesis over the other. Increase on lipid content might be accounted for increased lipid synthesis including the transformation of carbohydrate and protein into lipid, changes in the type of synthesis including decreased mobilization of lipid away from tissues. Increase in toxic concentration caused metabolic imbalance resulting in increased lipid level (Muley and Mane, 1989).

References

Barnes, H. and Stock, J. Black, 1973. Estimation of lipid in marine animal, detail investigation of the sulphophosphovanillin method. *J. Mar. Biol. Ecol.*, 12: 103–118.

Bhaktavathsalam and Reddy, S., 1983. Intoxication effect of lindane on the carbohydrate metabolism in the climbing perch, *Anabas testudineus. Pestic. Biochem. Physiol.*, 20: 340–346.

Ghosh, T.K., 1986. Nuvan induced physiological, biochemical and behavioral changes in *B. stigma. Poll. Res.*, 5: 3.

Kidd, H. and James, D. (Eds), 1994. *Agrochemicals Handbook*, 3rd Edn. Royal Society of Chemistry, Cambridge, England.

Mane, U.H. and Gokhale, A.A., 2002. Fluoride toxicity to the estuarine bivalve's molluscs. *J. Mar. Biol. Assoc., India*, 40(1–2): 16–29.

Meister, R.T. (Ed), 1995. *Farm Chemicals Handbook.* Meister Publishing Company, Willoughby, OH, p. 95.

Monoharan, T. and Subbiah, G.N., 1982. Toxic and sublethal effect of endosulfan on *Barbus stigma. Proc. Ind Acad. Sci.*, 92(6): 523–532.

Muley, D.V. and Mane, U.H., 1989. Endosulfan induced changes in the biochemical composition of the freshwater *Lamellibranch* mollusc, *Lamellidens marginalis* (Lamarck) from Godavari river, at Paithan. *Advances in Biosciences*, 8(1): 13–23.

Muley, D.V., 1991. Pesticide toxicity to freshwater lamellibranch molluscs from Maharashtra state, India. *Proc. 10th Int. Malacol. Cong.* (Tiibinger F.R.G., 1989), 207–211.

Nagabhushanam, R. and Dhamane, K.P., 1977. Seasonal variation in biochemical constituent of the clam, *Paphia laterisulca. Hydrobiologia*, 54(3): 209–214.

Rao, K.S.P. and Rao, K.V.R., 1981. Lipid derivatives in the tissue of the freshwater teleost, *S. mossambicus.* Effect of methyl parathion. *Proc. Ind. Nat. Acad.*, B47: 53–57.

Rao, K.S.P., Rao, K.R.S.S., Sahib, I.K.A. and Rao, K.V.R., 1985. Combined action of carbaryl and phenthoate on tissue lipid derivatives of mussel *Channa punctatus. Ecotoxicol. Environ. Saf.*, 9(1): 107–111.

Rao, L.M. and Ramaneswari, 2000. Effect of sublethal stress of endosufan and monocrotophos on the biochemical components of *L. rohita, M. vittatus* and *C. punctata. Ecol. Env. and Cons.*, 6(3): 289–296.

Saradamani and Kamalaveni, K., 2002. Sublethal effect of fenvalerate on bioenergetics and biochemical changes in *Cyprinus carpio. Indian J. Environ. and Ecoplan.*, 6(2): 315–318.

Swami, K.S., Rao, J.K.S., Reddy, S.K., Jagannatha Rao, K.K.S., Moorthy, L.C., Chetty, C.S. and Indira, S.K., 1983. The possible metabolic diversions adapted by the freshwater mussel to control the toxic metabolic effect of selected pesticides. *Ind. J. Comp. Anim. Physiol.*, 1: 95–103.

Chapter 42

Sodium Fluoride Induced Protein Alterations in Freshwater Fish, *Labeo rohita*

☆ ***M.D. Kale, S.A. Vhanalakar, S.S. Waghmode and D.V. Muley***

ABSTRACT

The toxicity of Sodium fluoride (NaF) to *Labeo rohita* was evaluated after the toxicant exposure to study the acute and chronic toxicity. Changes in protein content from selected tissues like muscle, liver, gill and kidney were recorded. Muscle shows the greatest loss of protein due to exposure of NaF followed by liver, gill and kidney.

Introduction

Inorganic Fluorides were introduced into the environment as a result of natural emission and anthropogenic sources. Depending on metrological condition and season, gaseous and particulate inorganic fluorides are transported in air and ultimately are deposited on land or open water bodies. Important anthropogenic sources of fluoride to the aquatic environment in clued municipal waste and effluents from fertilized producing plants and aluminum refineries (Woodiwiss and Fertwell, 1974). In water mobility and transport of inorganic fluoride are dependent on pH, water hardness, and the prescience of ion exchange mineral.

Inorganic fluoride are toxic to aquatic organism and may caused adverse biological effect such as change in carbohydrate, lipid, and protein metabolism, reproduction, impairment, reduce embryonic and development of life stage, and alternation size and growth (Samal, 1994).

Aquatic animals such as fish and invertebrates can take up fluoride directly from the water or via food (Hemens and Warwick, 1972; Nell and Livanos, 1988). Fluoride tends to be accumulated in the exoskeleton of invertebrates and in the bone tissue of fishes. Fluoride toxicity depends upon increasing fluoride concentration in the aquatic medium, exposure time and water temperature (Neuhold and Sigler, 1960; Angelovic *et al.*, 1961; Hemens and Warwick, 1972).

Sodium fluoride is the most common inorganic fluoride used in aquatic toxicity studies (Damkaer and Dey, 1989; Camargo, 1991). Sodium fluoride interrupts metabolic process such as glycolysis, lipid and protein synthesis in fishes (Camargo, 2003). Inorganic fluoride toxicity is negatively correlated to water hardness and positively correlated to temperature (Pimentel and Bulkley 1983).

Fluorine interferes with various metabolic activities and alters the levels of protein, lipids, glycogen, and cholesterol of fish (Kumar *et al.*, 2007). The present studies was under taken to evaluate the toxic effect on sodium fluoride on biochemical changes in different tissue such as gill, liver, kidney and muscle of freshwater carp *Labeo rohita*.

Material and Methods

The freshwater fishes *Labeo rohita* measuring about 6 to 7 cm in length were collected from state government fish seed rearing center. The collected fish were acclimatized under laboratory condition at 28-30°C for 10 days and then divided into different groups having 10 fishes in each. All the groups except control were transferred to separate plastic container containing different concentration (10 L) sodium fluoride (NaF) to determine toxicity LC_0 and LC_{50} value and fish behavior. Acute toxicity experiment was conducted for 96h and chronic toxicity for 30 days using a static bioassay technique. Toxic medium was changed at an interval of 24h. During experimentation temperature, pH, oxygen content and hardness of the water were determined using standard methods by APHA (1989). After acute exposure 96h fishes were sacrificed to obtained gills, liver, kidney and muscle. The pooled sample of the organ was used for estimation of total protein.

The experiment was conducted for acute and chronic exposure studies. After the completion of experimentation, the total protein was estimated by methods described by Lowry *et al.* (1951).

Result and Discussion

The alteration in total protein was calculated from *Labeo rohita* after acute and chronic exposure to sodium fluoride. The significant changes were observed in the experimental and control fish.

After the acute and chronic exposure to sodium fluoride, the protein content from all the tissues decreased significantly (Table 42.1 and Table 42.2). Muscle showed the greatest loss of protein as compared to all other tissues. In acute exposure studies, the muscle protein loss was more significant ($P<0.05$) followed by, liver, gill and kidney. The same pattern was observed after chronic exposure to sodium fluoride.

The initial phase of acute inorganic fluoride intoxication in freshwater species such as rainbow trout and carp is characterized by apathetic behavior. Fishes exposed to sodium fluoride become apathetic, loss weight, violent movement, increases secretion and wander aimlessly (Neuhold and Singler, 1960; Camargo and Tarazona, 1991). Sodium fluoride put impacts on aquatic animals by changing normal behavior at initial stage and alters the biochemical composition in later satge (Camargo, 2003). Exposure of Sodium fluoride to fish showed maximum loss of protein in muscle, followed by liver, gill and kidney. This decreased amount of protein may be due to the blocking of amino acid metabolism. The study showed that, sodium fluoride inhibits protein synthesis and interferes with amino acid metabolism (Pandit and Narayana, 1940). Another possible reason may be depletion of protein for its utilization in conversion to glucose (Sirvastava *et al.*, 2002).

Table 42.1: Changes in Protein Content in Different Tissue of *Labeo rohita* After Acute Exposure to Sodium Fluoride (96 hrs)

Tissue	*Control*	*Acute Dose*	
		LC_0	LC_{50}
Gill	17.73±0.33	12.43±0.57 *	9.73±0.64 **
Liver	21.90±0.20	16.85±1.00 *	13.83±0.85 **
Kidney	15.37±0.16	9.77±0.50 *	7.19±0.47 **
Muscle	23.45±0.24	15.22±0.46 *	10.02±0.62 **

Each value is the mean of three observations.

± S.D. values are significant at *: $P < 0.05$; **: $P< 0.01$; ***: $P< 0.001$.

Table 42.2: Changes in Protein Content in Different Tissue of *Labeo rohita* After Chronic Exposure to Sodium Fluoride (30 days)

Tissue	*Control*	*Chronic Dose*	
		LC_0	LC_{50}
Gill	15.52±0.22	11.16±0.46 **	7.27±0.49 ***
Liver	17.00±0.29	12.90±0.68 **	8.13±0.54 ***
Kidney	13.54±0.74	10.29±0.64 **	7.09±0.46 ***
Muscle	18.43±0.59	11.22±0.47 **	7.73±0.51 ***

Each value is the mean of three observations.

± S.D. values are significant at *: $P < 0.05$; **: $P< 0.01$; ***: $P< 0.001$.

Significant alternation in protein metabolism on acetylcholinesterase activities and oxygen consumption in freshwater crabs have been described by under fluoride toxication (Reddy and Venugopal, 1990). The effect caused by exposure to inorganic fluoride has been also observed in aquatic animals by Sigler and Newhold, 1972; Mishra and Mohapatra 1998. Kumar *et al.* (2007) showed the significant decrease in protein by exposing freshwater catfish (Clarias batrachus, Linn.).

References

Angelovic, J.W., Sigler, W.F. and Neuhold, J.M., 1961. Temperature and fluorosis in rainbow trout. *J. Water Pollut. Control Fed.*, pp. 371–381.

APHA, 1989. *Standard Methods for the Examination of Water and Wastewater*, 17th Edn. American Public Health Association, Washington.

Camargo, J.A., 2003. Fluoride toxicity to aquatic organism: A review. *Chemosphere*, 50(3): 251–264.

Camargo, J.A., 1991. Eco-toxicological analysis of the influence of an industrial effluent on fish population in a regulated stream. *Aquacult. and Fish. Manage.*, 22: 509–518.

Camargo, J.A. and Tarazona, J.V., 1991. Short term toxicity of fluoride ion (F^-) in soft water to rainbow and brown trout. *Chemosphere*, 22: 605–611.

Damkaer, D.M. and Dey, D.B., 1989. Evidence for fluoride effect on salmon passage at John Day Dam, Columbia River, 1982–1986. *N. Am. J. Fish. Managt.*, 9(2): 154–162.

Hemens, J. and Warwick, R.J., 1972. The effects of fluoride on estuarine organisms. *Water Res.*, 6: 1301–1308.

Kumar, A., Tripathi, N. and Tripathi, M., 2007. Fluoride induced biochemical changes in freshwater catfish (*Clarias batrachus*, Linn.). *Fluoride*, 40(1): 37–41.

Lowry, O.H., Rosebrough, N.J., Farr, A.B. and Randall, R.J., 1951. Protein measurement with folin-phenol reagent. *J. Bio. Chem.*, 193: 265–275.

Mishra, P.C. and Mohapatra, A.K., 1998. Haematological characteristics and bone fluoride content in *Bufo melanostictus* from an aluminum industrial site. *Enviorn. Pollut.*, 99(3): 421–423.

Nell, J.A. and Livanos, G., 1988. Effects of fluoride concentration in seawater on growth and fluoride accumulation by Sydney rock oyster (*Saccostrea commercialis*) and flat oyster (*Ostrea angasi*) spat. *Water Res.*, 22: 749–753.

Neuhold, J.M. and Sigler, W.F., 1960. Effect of sodium fluoride on carp and rainbow trout. *Trans. Am. Fish. Doc.*, 89: 358–370.

Pandit, C.G. and Narayana, R.D., 1940. Endemic fluorosis in south India: Experimental production of chronic fluorine intoxication in monkey (*Macaca radiate*). *Ind. J. Med. Res.*, 28: 559–574.

Pimentel, R. and Bulkley, R.V., 1983. Influence of water hardness of fluoride toxicity to rainbow trout. *Environ. Toxicol. Chem.*, 2: 381–386.

Reddy, S.L.N. and Venugopal, N.B.R.K., 1990. Effect of fluoride on acetylcholinesterase activity and oxygen consumption in a freshwater field crab, *Barytelphusa guerini. Bull. Environ. Cont. and Toxicol.*, 760–766.

Samal, U.N., 1994. Effect of fluoride on growth of certain freshwater fishes. *Environ. Ecol.*, 12(1): 218–220.

Sashi, Singh J.P. and Thaper, S.P., 1989. Effect of fluoride in excess on lipid constituents of respiratory organs in albino rabbits. *Fluoride*, 22(1): 33–39.

Sigler, W.F. and Neuhold, J.M., 1972. Fluoride intoxication in fish: A review. *J. Wild. Dis.*, 8: 252–254.

Sirvastava, N., Kaushik, N. and Gupta, P., 2002. Zink induced changes in the liver and muscle of fish *Channa punctatus* (Bloch). *J. Ecophysiol. Occup. Hlth.*, p. 197–204.

Woodiwiss, F.S. and Fertwell, G., 1974. The toxicity of sewage effluents, industrial discharge and some chemical substance to brown trout (*Salmo trutta*). In the Trent River authority area. *Water Pollut. Cont. (G.B.)*, 73: 396.

Index